VISION

HUMAN AND ELECTRONIC

OPTICAL PHYSICS AND ENGINEERING

Series Editor: William L. Wolfe

Optical Sciences Center
University of Arizona
Tucson, Arizona

1968:
M. A. Bramson
Infrared Radiation: A Handbook for Applications

1969:
Sol Nudelman and S. S. Mitra, Editors
Optical Properties of Solids

1970:
S. S. Mitra and Sol Nudelman, Editors
Far-Infrared Properties of Solids

1971:
Lucien M. Biberman and Sol Nudelman, Editors
Photoelectronic Imaging Devices
  Volume 1: Physical Processes and Methods of Analysis
  Volume 2: Devices and Their Evaluation

1972:
A. M. Ratner
Spectral, Spatial, and Temporal Properties of Lasers

1973:
Lucien M. Biberman, Editor
Perception of Displayed Information

W. B. Allan
Fibre Optics: Theory and Practice

Albert Rose
Vision: Human and Electronic

VISION
HUMAN AND ELECTRONIC

Albert Rose

David Sarnoff Research Center
RCA
Princeton, New Jersey

PLENUM PRESS • NEW YORK-LONDON

Library of Congress Cataloging in Publication Data

Rose, Albert, 1910–
Vision: human and electronic.

(Optical physics and engineering)
Includes bibliographies.
1. Photoelectronic devices. 2. Vision. 3. Television picture tubes. 4. Photography–Films. I. Title. [DNLM: 1. Optics. 2. Vision. 3. Visual perception. WW103 R795v 1973]
TK830.R67 621.3815'42 73-97422
ISBN 0-306-30732-4

First Printing - January 1974
Second Printing - November 1974

A Division of Plenum Publishing Corporation
227 West 17th Street, New York, N.Y. 10011

United Kingdom edition published by Plenum Press, London
A Divsion of Plenum Publishing Company, Ltd.
Davis House (4th Floor), 8 Scrubs Lane, Harlesden, London, NW10 6SE, England

Printed in the United States of America

To Lillian

PREFACE

The content of this monograph stems from the writer's early involvement with the design of a series of television camera tubes: the orthicon, the image orthicon and the vidicon. These tubes and their variations, have, at different times been the "eyes" of the television system almost from its inception in 1939. It was natural, during the course of this work, to have a parallel interest in the human visual system as well as in the silver halide photographic process. The problem facing the television system was the same as that facing the human visual and the photographic systems, namely, to abstract the maximum amount of information out of a limited quantity of light. The human eye and photographic film both represented advanced states of development and both surpassed, in their performance, the early efforts on television camera tubes. It was particularly true and "plain to see" that each improvement and refinement of the television camera only served to accentuate the remarkable design of the human eye. A succession of radical advances in camera-tube sensitivity found the eye still operating at levels of illumination too low for the television camera tube. It is only recently that the television camera tube has finally matched and even somewhat exceeded the performance of the human eye at low light levels.

It was also clear throughout the work on television camera tubes that the final goal of any visual system—biological, chemical, or electronic—was the ability to detect or count individual photons.

A finite quantity of light meant a finite number of photons, and a finite number of photons meant a finite amount of information. Only by counting every photon could the total information in the light flux be extracted. These elementary statements define the emphasis of this monograph.

The emphasis is on several major aspects of photon counting and can be put in the form of questions. What is the information content of a finite number of photons? How well do the several systems—biological, chemical, and electronic—achieve the goal of photon counting? And finally, what is the underlying physics governing the counting processes in electronic systems?

The emphasis is, in brief, on the sensitivity performance of the several systems as measured on an absolute scale. The absolute scale is the ratio of information transmitted by the particular visual system to the information content of the light flux entering the system.

The present monograph is not intended in any way as a complete discussion of the state of the art on human vision or on the various electronic devices—except in the one respect of the achievement of the absolute limit of sensitivity. There is obviously a wide variety of visual phenomena such as color vision, the structure and biochemistry of the retina and the organization of the optic nerve fibers which is touched only tangentially. Similarly, there is an extensive literature concerning the design, processing, and performance of particular electronic systems, much of which is reviewed in the two volumes edited by Lucien N. Biberman and Sol Nudelman entitled *Photoelectronic Imaging Devices*.* This literature is referred to only as it illuminates the central theme of the present monograph.

The literature on human vision, photography, and television normally appears in widely separated journals. It is hoped that the present treatment of these divergent fields in terms of a common theme will contribute to the essential process of cross-fertilization. To that end, considerable emphasis has been placed on the presentation in elementary physical terms of a number of topics whose highly formalized literature has been accessible only to small groups of experts. These topics include noise currents, gain-bandwidth

* Published by Plenum Press, New York (1971).

limitations in solid-state devices, and electron-phonon interactions. The obviously interdisciplinary character of the monograph may, indeed, extend even to several fields within the discipline of solid-state electronics.

This monograph incorporates many of the writer's professional interests over a period of some decades during which time he has enjoyed the stimulation and guidance of a generous number of colleagues. I am indebted to Harley Iams, Paul Weimer, Harold Law, A. Danforth Cope, and Stanley Forgue for the adventure of exploring new forms of television camera tubes; to Dwight O. North for numerous discussions of stochastic processes; to Otto Schade for his catholic understanding of and contributions to almost all aspects of television, photographic, and human visual systems; to Roland Smith and Richard Bube for the opportunity of interacting with their early work on photoconductivity; to Wolfgang Ruppel, Henry Gerritsen, Fritz Stockmann, and James Amick for a cooperative effort on understanding the physics of electrophotographic systems; to Murray Lampert for an extended and deeply rewarding collaboration on injection current in solids; to Richard Williams, Helmut Kiess, and Avram Many for their penetrating insights into high field transport; and to George Whitfield, Allen Rothwarf, and Lionel Friedman for expert guidance on electron-phonon interactions. I am also indebted to R. Clark Jones for his generous support and elaboration of my early arguments for the quantum limitations of human vision. I would like to thank A. Many, W. Low, and M. Schrieber of the Hebrew University for the opportunity of exploring the coherence of the subject matter of this monograph in a series of lectures for the Bathsheva de Rothschild Seminar on Applied Physics. Finally, I have profited from a critical reading of the manuscript by Ralph Engstrom, Helmut Kiess, and Otto Schade.

The final writing of the manuscript was carried out in the stimulating and congenial atmosphere of Laboratories RCA Ltd in Zurich, for which I am indebted to the director, Walter Merz, and the staff.

CONTENTS

UNITS AND DEFINITIONS

Photometric Units

1 lumen of green light	=	1.5×10^{-3} watt
	=	4×10^{15} photons/sec
1 lumen of white light	≐	4×10^{-3} watt
	≐	10^{16} photons/sec

Incident Illumination

1 lumen/m^2	≡	1 lux	≡ 1 meter-candle
1 lumen/ft^2	≡	1 foot-candle	≐ 10 lux

Surface Brightness

1 lumen/ft^2 falling on a 100% reflecting and perfectly-diffusing surface yields a surface brightness of 1 foot-lambert. Similarly, 1 lumen/cm^2 yields a surface brightness of 1 lambert.

1 foot-lambert ≐ 1 millilambert

CHAPTER 1

THE VISUAL PROCESS

1.1. Introduction

It would be difficult to find a more cogent confrontation between physics and biology than in the visual process. Nature was faced from the beginning with the hard fact that light consists of a finite number of bits of energy, called "photons" or "quanta." Whatever visual information was to be distilled out of the surrounding world was circumscribed by the profound constraints imposed by the discrete nature of light.

Throughout the millions of years over which life in its manifold forms evolved, survival was the dominant motif. Unless the prey could detect its predator in ample time, life terminated abruptly. Visual detection was not the only means of detection, but it was a major one. And visual detection had still to function in twilight and even in starlight when the stream of photons dwindled to an occasional patter of drops of energy. It was, indeed, a matter of life and death that each photon be husbanded and assembled to trace out the best possible image of impending disaster. Little short of a photon counter would suffice.

There is ample evidence among the primitive forms of life that nature mastered the art of counting photons at an early age. If it were only a question of utilizing the incident energy of the photons, we can, in fact, predate animal life and point to the highly efficient solar battery developed in plants by way of photosynthesis. But the counting of photons entails not only the efficient absorption of photons but also the highly sophisticated process of amplification. The energy of a photon is sufficient to disturb only a single atom or molecule. With this energy alone, the information that a photon had been absorbed could not be transmitted beyond the point of absorption, let alone to some central nervous system. A nerve pulse involves the motion of at least some millions of atoms or ions. Hence, the energy of the absorbed photon must be multiplied or amplified over a millionfold before it can give rise to a nerve pulse. The ingenious amplifier that nature devised remains an unsolved puzzle. The great variety of amplifiers that man has devised for the same purpose is in large part the content of this monograph.

The quantum character of light is a hard constraint. Nature could, in a physical sense, do no more and, in a survival sense, do no less than devise a photon counter. Once having the photon counter, there were secondary choices as to how the information was to be handled.

The incoming photons, for example, could be accumulated for a long time to generate an image of high quality or for short times to give a rapid series of low-quality images. The long-time accumulation would mean that moving objects would be blurred. Moreover, the animal itself would have to slow down its movements so that it did not joggle its camera (or visual system) during the course of an exposure. At the other extreme, a radically shortened time of exposure would yield images of such poor quality or so impoverished of information content as to be of little value in guiding the animal's response. The compromise was set, at least for the human system, where one might expect, namely, at an exposure time matched to the reaction time of the human system as a whole. The reaction time is the sum of the transit time of nerve pulses from eye to brain and back to an appropriate extremity plus the time required to overcome the physical inertia of that extremity. Overall, the reaction time is in the order of a tenth of a second, as is the exposure time of the eye.

The choice of exposure time is readily understandable. So also is the choice of spectral response. The latter peaks near the peak of the sun's radiation—and even shifts at twilight toward the blue in order to match the shifting spectral content of the light scattered from the "blue" sky. There are a host of other choices perhaps not so obvious and perhaps offering the opportunity for reading the past through the shape of the present. These have to do with the size of lens opening, the focal length of the lens, the red and the blue cutoffs of the visible range, the density of retinal elements which puts a ceiling on image quality, the multiplicity of color vision, the programed interconnections of the optic nerve fibers, and, finally, even the number and location of eyes. We will return to this subject later and point out some of the adaptations of the optical parameters to the life habits of a number of animals. In the meantime, we note the contrast between the primary character of the photon-counting problem—it is singular, ultimate, and essential—and the wide-ranging secondary character of the ways in which the photon counting was adapted to the life habits of particular animals.

1.2. Quantum Limitations on the Visual Process

The absolute measure of the performance of a visual system is the ratio of information transmitted by the system to the information incident on the system and contained in the incident light flux. It is necessary, then, that we have a quantitative measure of the information conveyed by a finite number of photons. We derive such a measure in a series of steps designed to emphasize three aspects of the quantum limitations on the visual process. The first aspect has to do with the overall finite number of photons; the second aspect is their random distribution in time and space; and the third aspect is the problem of guarding against false alarms, that is, spurious visual patterns that may arise from the random character of the photon distribution and not from the original scene itself. Each of these aspects exacts an increasing toll in terms of the number of photons required to transmit an elementary bit of information.

1.2.1. Discreteness of Light Quanta

We imagine first that we have a black canvas and that we wish to paint a picture (Fig. 1.1) on the canvas depicting a white wall on which is located a single black spot. This is the simplest

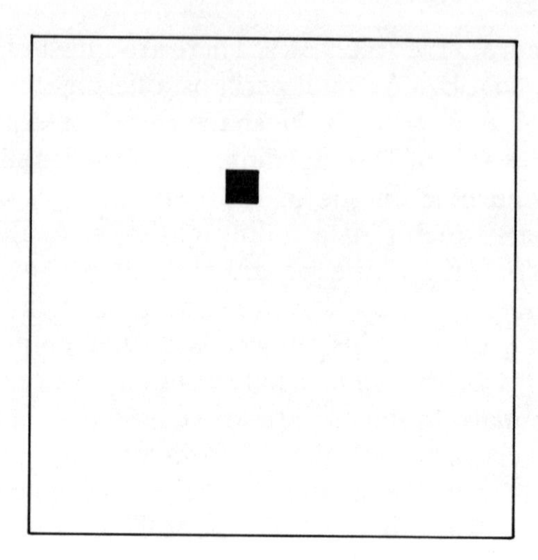

Fig. 1.1. Test pattern consisting of a single black spot on a white background.

of pictures in that we wish to indicate only the presence of the black spot, and not its structure, on the otherwise white wall. Furthermore, the method of painting will be constrained to be that of stippling. We can paint an array of small white dots, all of the same size but of varying spacing. Each white dot will correspond to the visual effect of a photon in a generalized visual system.

We suppose that the size of the black spot is such that the canvas can accommodate a total of N of these close-packed spots. The single black spot, then, defines the size of a picture element whose area is a fraction N^{-1} of the canvas.

At this point, we ask what would be the smallest number of white dots required to portray the presence of one black spot on a uniformly white wall? If we are allowed to space the white dots uniformly, then, clearly, $N - 1$ white dots are both necessary and sufficient to complete the canvas. The single missing white dot locates the presence of a single black spot (Fig. 1.2).

The next step in sophistication of our painting will be to locate a single gray spot as well as to portray its shade of grayness. We assume that the reflectivity of the spot is 99% of that of the

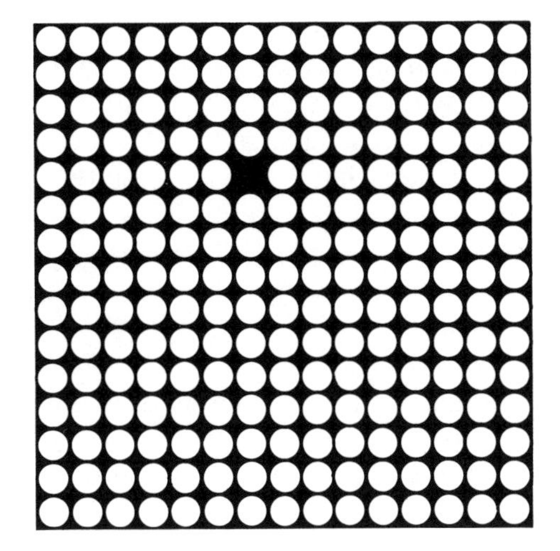

Fig. 1.2. Reproduction of Fig. 1.1 using an ordered array of "photons."

white wall and again ask for the smallest number of white dots required to convey this information. The answer, obviously, is $100N - 1$ dots. Each picture element will have exactly 100 dots stippled in, with the exception of the one picture element containing the gray spot. The latter will have 99 dots to indicate that its brightness is 99% of that of the surrounding wall.

While all of the above is embarrassingly elementary, the argument does stress the high cost in photons required to portray small elements of low contrast. For example, the number of picture elements N required for well-resolved images often lies in the range of 10^6–10^7. Hence, we would need some 10^8–10^9 photons to delineate the location and brightness of the gray spot.* That is,

* While a single gray spot on a white wall may in one sense appear to transmit only a single item of information, the total picture transmits as much information as any more complex pattern. The picture contains the information that there is *no* gray spot on any of the other $N - 1$ picture elements. Hence, the brightness of each picture element is a discrete independent item of information even when the picture is a complete "blank."

we would need these $100N$ photons providing they could be arranged in a precise array of 100 photons per picture element. But nature does not work in so orderly a fashion. Photons arrive at random times and places and give rise to a fundamental graininess in any image, a graininess that tends to obscure the detection of fine detail and faint contrasts. The result is a considerable increase in the number of photons required to delineate the fine detail of images.

1.2.2. Random Character of Photon Distributions

Natural incoherent light is emitted by some form of electronic excitation, as from an excited atom. The average lifetime of the excited state is a well-defined, calculable, and observable parameter. On the other hand, it is a fundamental property of the quantum-mechanical nature of such an excitation that the photon can be emitted at any time during the average life of the excited state. More definitely, the *probability* of emitting a photon at any time t and in a time interval Δt is given by $\exp(-t/\tau)\,\Delta t/\tau$, where τ is the average lifetime of the excited state. What is significant for our purposes, is that the emission of photons is a stochastic process.

If we carry out the experiment of illuminating a small area with a "constant" incandescent light source and count the number of photons that strike the area in a given time Δt, we will obtain a series of numbers $n_1, n_2, \ldots$ corresponding to the actual number of photons that arrived at the first interval Δt, the second interval Δt, and so on. We have put quotation marks around the word "constant" because the experiment which we are performing is one way of determining whether the source is indeed constant. The fact that the numbers of the series $n_1, n_2, \ldots$ do not all have the same value may in the broadest sense cast doubt upon the constancy of the source. Yet, no matter how carefully we design the source, it will turn out that there is an irreducible spread to these numbers. That spread is the consequence of the stochastic or random nature of the process of emitting photons.

In particular, it will be true that if we observe that the average number of photons arriving at the test area is n_0, we will also find that the numbers $n_1, n_2, \ldots$ are distributed around n_0 in such a fashion that the average value of $(n_i - n_0)^2$ will also be

n_0. The average value of $(n_i - n_0)^2$ is called the mean-squared deviation from the mean. The square root of this quantity, or $\langle (n_i - n_0)^2 \rangle^{1/2}$ is called the *root mean squared deviation* from the mean and is abbreviated as *rms deviation.*

We introduce here also a terminology that will be used frequently in this monograph. A *signal* is defined as the *average* number of photons falling on a test element. The *noise* is the *rms deviation* from this number. In the example cited above, n_0 is the value of the signal and $n_0^{1/2}$ is the value of the noise. The signal-to-noise ratio is then also $n_0^{1/2}$. The term signal will also be used, as will be clear from the context, to mean the *difference* between the average numbers of photons falling on a given test element and on surrounding test elements of the same size. This meaning is used, for example, to define the signal appropriate to a low-contrast test-element on a uniform surround.

We return to the black canvas on which we wish to paint a small gray spot by use of a stippling of white dots, each representing a photon. In our first estimate, using uniformly spaced dots, we arrived at the need for 100 dots per picture element (a picture element was defined as the area of the gray spot to be portrayed) in order to portray a single gray spot having 99 % of the brightness of the surrounding canvas. If we now recognize the random character of the photon distribution, we will find that the actual numbers of photons falling on various picture-element areas are distributed around the average number 100 such that the rms deviation is $(100)^{1/2}$ or 10.* At this point, we have a signal to be detected which is 1 % of the surrounding average brightness and we are faced with a noise fluctuation in these picture element areas which is 10 % of the average brightness. (The signal to be detected in this example is the difference between the number of photons in the test element and the average number of photons in equal area elements of the surround.) In brief, the signal-to-noise ratio is 0.1, and far less than the value of unity which is frequently taken as the threshold for visibility of a signal against a noisy background.

* The rms deviation will be the same whether we look at the numbers of photons falling on a given area in successive equal time intervals or at the numbers of photons falling on many equal areas in a single time interval.

Our first requirement, then, is to increase the density of photons on the canvas so that the fluctuation or noise level does not exceed the signal to be detected. Since the signal represents a 1% deviation from the surround, we require that the noise level (or rms deviation) also not exceed 1%. This is achieved by having an average of 10^4 photons falling on each picture element. The rms deviation will then be the square root of 10^4, or 10^2. And the ratio of this random deviation to the average will be 10^{-2}, or 1%.

In summary, at this point, the number of photons required to portray a single spot was increased by a factor of 100 in going from a black spot to a gray spot having only 1% contrast with the surround; and the number was increased again by a factor of 100 in going from an ordered array of photons to a random array. The latter factor insured that the signal to be detected was equal to the rms deviations occurring as a result of the fundamentally random character of the photons. There is yet another another factor to be introduced to guard against false alarms, that is, the mistaking of any particular random fluctuation for the real signal to be detected.

1.2.3. False Alarms

It is frequently stated or implied in discussions of electronic systems that the threshold of detectability of a signal occurs when the signal is equal to the noise. This is a somewhat misleading statement. For example, suppose that we are monitoring the level of an electrical current to detect significant changes of 1% or more. Suppose, also, that the noise level (rms deviation) is 1% of the average current. If we make N successive observations of the current, we will find that the current will depart from its average value by 1% or more in almost half of the observations even in the absence of any "real" or deliberately imposed disturbance of the current. We must put quotes on the term "real" because the fluctuations resulting from the noise are just as "real" as a deliberately imposed disturbance—it is only that the source of the disturbance is different. The significant part is that almost half of our observations will tell us that the current has departed from its average value by more than 1% whether or not there has been any deliberate disturbance. It is in this sense that half the observations will be false alarms. In order to guard against false alarms, the "real" signal to be detected must exceed the level of noise by some appropriate factor. The factor can readily be approximated by knowing the statistical distribution of noise fluctuations as well as the number of observations which should statistically be free from false alarms.

Figure 1.3 shows the distribution of noise fluctuations around the mean value of a parameter. The ordinate is the probability density and the abscissa k is plotted in units of the rms deviation. The second abscissa scale, n, is a particular numerical example for which the average number of photons is 900. The rms deviation is then 30. The total area under the curve using the k abscissa scale is unity. The area under the curve between $k = 1$ and $k = 2$, for example, is 0.13 and represents the probability that an observation will lie in the range between 1 and 2 rms deviations above the mean. In the numerical example, it is the probability that an observation will lie between 930 and 960 photons. Similarly, the

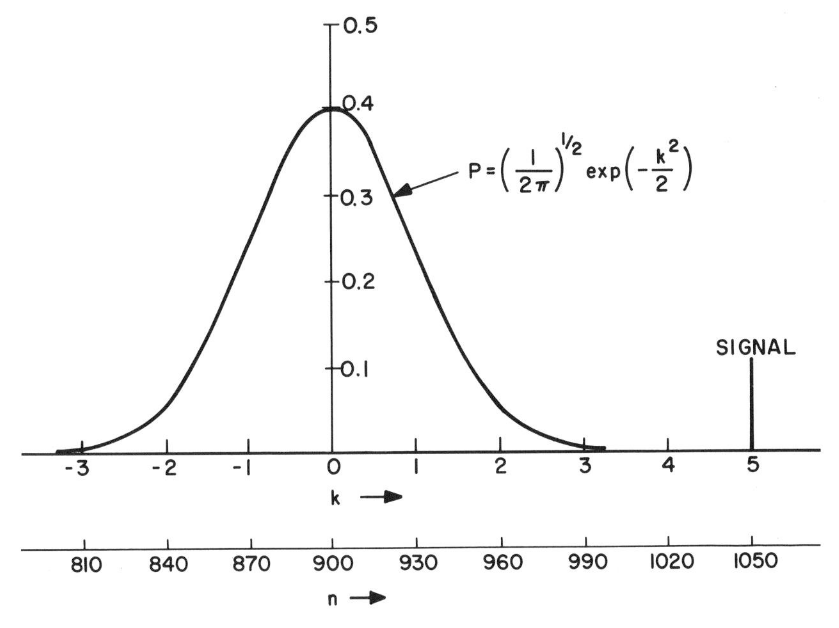

Fig. 1.3. Probability distribution of a noisy quantity about its mean value.

area under the curve to the right of $k = 2$ is 0.023 and represents the probability that an observation will exceed 2 rms deviation units above the mean. In the numerical example, it is the probability that an observation will exceed 960 photons.

Table 1.1. gives the probability that noise fluctuations will exceed the mean value of the background by 1, 2, 3, etc. units of the rms value of the noise. The probability for noise fluctuations occurring on either side of the mean is just twice the probabilities listed.

With the aid of Table 1.1 we can now specify how large a signal is required in order to avoid false alarms. The signal in this case is the difference in average brightness between the test spot and the background. We suppose, as is common, that the picture has 10^5 picture elements, each of the size or area occupied by the test spot. We have then 10^5 opportunities to generate a false alarm. And, if our purpose is to reduce the number of false alarms to below unity, we will need, according to Table 1.1, a signal whose amplitude is 4–5 times larger than the rms noise. We call this value of k the *threshold signal-to-noise ratio.* It is such a value of signal for which we are reasonably confident of not mistaking a noise fluctuation for the real signal. Note that at $k = 6$ the probability of detecting a false alarm is already far smaller than is needed. In the other direction, $k = 3$ would only guard against false alarms for a picture having less than 10^3 picture elements. Hence $k = 5$ is a reasonable approximation to the threshold signal-to-noise ratio.

Table 1.1
Values for the Probability of Exceeding Various Values of k

k	Probability of exceeding k
1	0.15
2	0.023
3	1.3×10^{-3}
4	3×10^{-5}
5	3×10^{-7}
6	2×10^{-9}

We choose $k = 5$ rather than $k = 4$ for the following reason. In the above argument, we assumed a very noisy background and a well-defined signal. However, the signal itself has nearly the same noise or spread as the background.* This means, for example, that if in Fig. 1.3 we located the mean value of the signal at $k = 4$, we would find that the signal appeared half the time below $k = 4$, and half the time above. If, on the other hand, we locate the signal at $k = 5$, we will find that only 0.15 of the time will the signal appear below $k = 4$. It will exceed $k = 4$, on the average, 0.85 of the time and be judged a real signal. Hence, a margin of about one unit of k above the nominal value needed to avoid false alarms is sufficient to give a reasonable reliability to our observations.

At this point, we compute the increase in photon density required to satisfy the criterion $k = 5$, as compared with the criterion $k = 1$ used in the previous section. We begin with the condition $k = 1$ for which the signal is equal to the rms deviation of the noise. In particular, let the signal and the rms deviation each be 1% of the background brightness. As we increase the photon density, the signal remains constant when measured as a percentage of the background brightness. The rms deviation of the noise, however, decreases. Since the ratio of the rms deviation to the average background brightness varies as $n_0^{1/2}/n_0 = 1/n_0^{1/2}$, where n_0 is the average density of photons in the background, it will be necessary to increase n_0 by a factor of k^2 $(= 25)$ in order to decrease the ratio $1/n_0^{1/2}$ by k $(=5)$.

In sum, the density of photons required varies as k^2. And, for the value $k = 5$, the density of photons must be increased 25-fold relative to the density computed in the previous section for $k = 1$. In the previous section, the number of photons was computed to be $10^4 N$, where N was the number of picture elements. Hence, to guard against false alarms, this number must be increased to $2.5 \times 10^5 N$.

We can now write, in general, the expression for the total number of photons required to detect a contrast C where C is a

* Black or very dark signal elements on a white background must, of course, be excepted.

measure of the signal as a fraction of the background brightness, that is, $C \equiv \Delta B/B$ and $0 \leqq C \leqq 1$ ($C = 1$ means 100% contrast and $C = 0.01$ means 1% contrast).

$$\text{Total number of photons} = N \frac{1}{C^2} k^2 \tag{1.1}$$

Here, N is the total number of picture elements and reflects the discreteness of the photons. The factor $1/C^2$ is a consequence of the contrast C and the random character of photon distributions; the factor k^2 reflects both the random character of the photon distribution and the need to avoid false alarms.

1.3. A Summary Experiment

Almost all of the conclusions of the previous three sections can be read off by inspection of Fig. 1.4. In Fig. 1.4a there is depicted an area uniformly illuminated by a low density of photons. Each photon was made visible on a television screen as a discrete white dot by using a high-gain photomultiplier. We note in Fig. 1.4a the discrete character of photons, their random distribution, and their consequent noisiness which gives rise to false alarms. A simple inspection of Fig. 1.4a reveals black areas or spots which we could, in the absence of other information, readily identify as "real" black spots in the original picture. In fact, no such deliberate pattern of black spots was introduced into the making of Fig. 1.4a. The black spots are a consequence of the statistical fluctuations in the distribution of photons.

Figure 1.4b is a "real" test pattern of black spots* which we wish to detect under the low illumination represented by Fig. 1.4a. To do so we simply superimposed the positive transparencies of Figs. 1.4a and 1.4b to obtain Fig. 1.4c. Note that the four larger black spots of Fig. 1.4b are readily visible in Fig. 1.4c. The remaining smaller black spots of Fig. 1.4b are undetectable in Fig. 1.4c. They are lost in the noise. Note also that the four larger spots that are visible in Fig. 1.4c *appear* to terminate in a fifth black spot forming the apex of a triangle. The fifth black spot, however,

* The thin black bar will be referred to in Section 1.5.

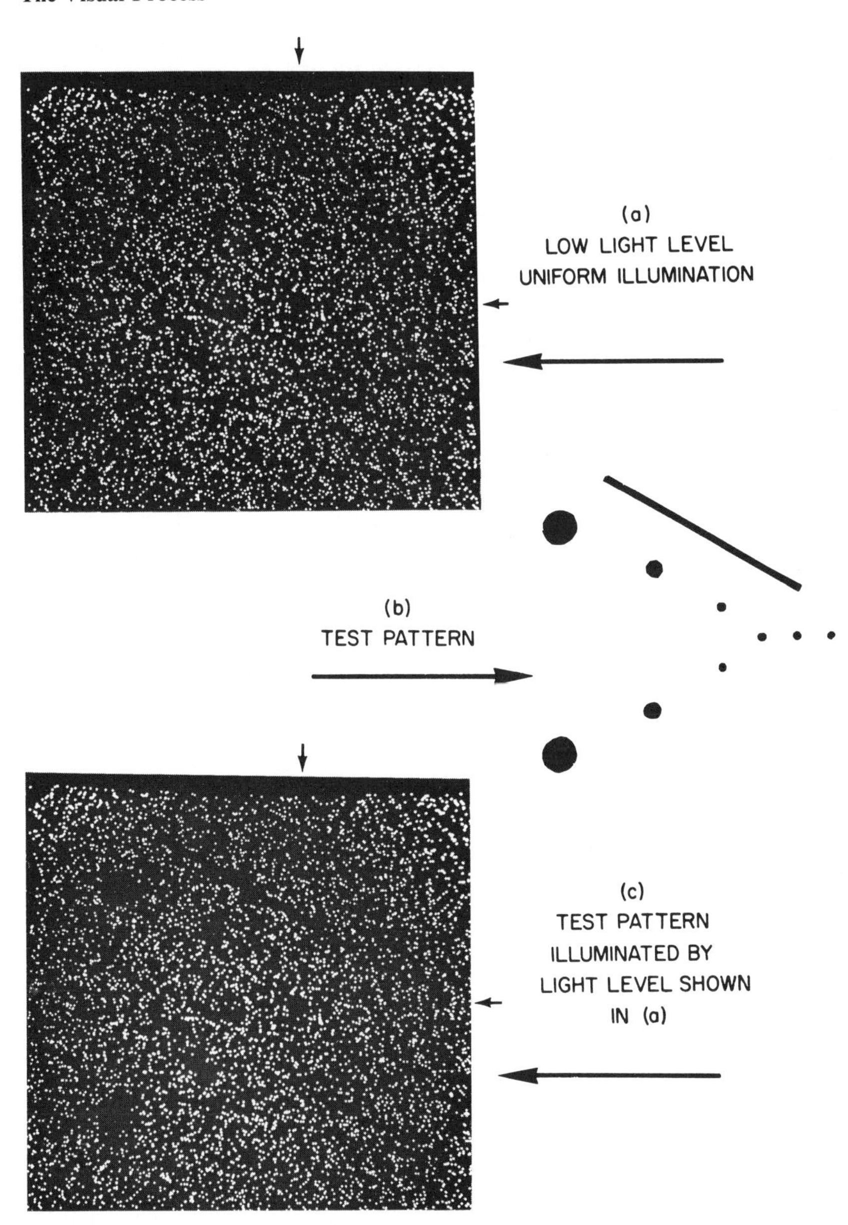

Fig. 1.4. Demonstration of the limitations imposed by the quantum nature of light on its ability to transmit information.

is one of the largest statistically generated black spots *already* present in the "uniform" illumination of Fig. 1.4a. For convenience, this black spot is located in Figs. 1.4a and 1.4c by small coordinate arrows on the edges of the pictures.

The presence of this statistically generated black spot in Figs. 1.4a and 1.4c means that any "real" black spot must be larger than it in order to be reliably judged to be "real." If we take this statistically generated black spot as a starting point, we find that it occupies about 1/500 of the area of the picture. Also, since there are some 4500 dots in the picture, the average number of dots in the area of this spot is 9. The signal-to-noise ratio is then $\sqrt{9} = 3$. From Table 1.1, we note that a signal-to-noise ratio of 3 would yield false alarms only one in a thousand times. Hence, since there are 500 picture elements having the size of the black spot, and since the probability of obtaining such a black spot by statistical fluctuation is one part in a thousand, we are at the threshold of reliable visibility, of "real" signals or "real" black spots. The "real" black spots need only be somewhat larger than the one we are considering. In particular if the signal-to-noise ratio of the "real" spot were between 3 and 4, we would have sufficient confidence in its reality. Note, for example, that the signal-to-noise ratio of the largest black spots in Fig. 1.4c is approximately 5, since they each obscure about 25 white dots. These large black spots are clearly well above the threshold of reliable visibility and tend to confirm that our estimate of threshold signal-to-noise ratio should lie between 3 and 4.

The density of photons in Fig. 1.4 is at the extreme low end of densities that we normally encounter. The low density was deliberately chosen to illustrate the three major properties of photon distributions: discreteness, random distribution, and false alarms. In the range of the higher densities normally encountered, the number of picture elements is likely to be of the order of 10^6 rather than the 10^3 calculated for Fig. 1.4. Under these conditions, the threshold signal-to-noise ratio also must be increased to values between 4 and 5 in order to guard against the appearance of false alarms. From Table 1.1., at $k = 5$, the probability of false alarms is only 3×10^{-7}. It decreases rapidly at $k = 6$ to 2×10^{-9}. A television picture, for example, has some 10^5 picture elements and would call for a value of k between 4 and 5.

Figure 1.4a serves another purpose. It emphasizes how completely unrealistic it would be to use the customary criterion for visibility, namely, a signal-to-noise ratio of unity. The operational meaning of this criterion is that if a first person removed one of the dots in Fig. 1.4a, a second person could determine which dot had been removed. The removal of one dot is the equivalent of having a black spot in a test pattern that obscures, on the average, one dot or photon. The average signal is then one photon and the noise, which is the square root of the average, is also one photon, yielding a signal-to-noise ratio of unity. Simple inspection of Fig. 1.4 shows the virtual impossibility of detecting a single missing photon.

1.4. A Second Experiment

The experiment of Fig. 1.4 was confined to the visibility of black spots for which the contrast is, by definition, unity. If we choose to look for gray spots for which the contrast is less than unity, the size of the spot must be increased. According to Eq. (1.1),

$$\begin{aligned}\text{Total number of photons} &= N \frac{1}{C^2} k^2 \\ &= \frac{A}{d^2 C^2} k^2 \qquad (1.2)\end{aligned}$$

where d is the linear dimension of a picture element (that is, the test spot) and A is the area of the picture. Equation (1.2) states that if we keep the total number of photons fixed (that is, maintain constant brightness), the diameter d of a test spot which is just visible should vary inversely with its contrast C.

Figure 1.5 is a photograph of a test pattern made up of discs whose diameters decrease by a factor of 2 in progressing along a row, and whose contrasts decrease by a factor of 2 in progressing down a column. Along a 45° diagonal of Fig. 1.5, the product dC is then constant. If we illuminate Fig. 1.5 with some intermediate light intensity, the boundary between the discernible and non-discernible parts of Fig. 1.5 should be one of the 45° diagonals.

Figure 1.6 shows the results of a series of illuminations of Fig. 1.5. Figure 1.6 was obtained by photographing the kinescope of a television system when the pattern of Fig. 1.5 was illuminated

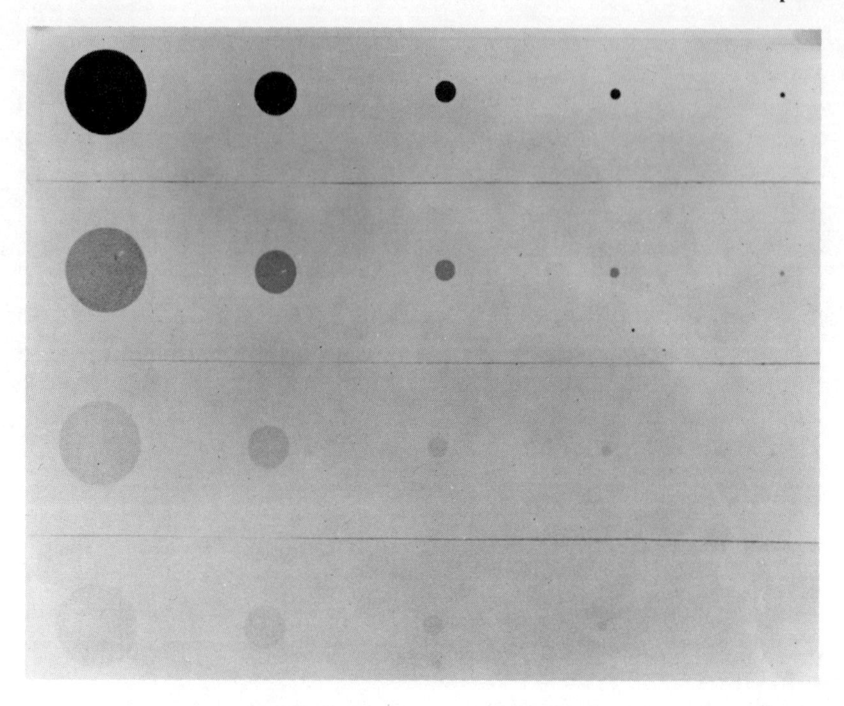

Fig. 1.5. Test pattern used to measure the resolving power of a system in terms of the size and contrast of single elements.

by a flying-spot scanner and the reflected light recorded by a high-gain photomultiplier.[R-1] Parenthetically, Fig. 1.6 was made some years prior to Fig. 1.4 and shows a certain variation in the intensities of the individual white dots, each being the trace or signal of an individual photon. The variation arises from the variation of gain of a photomultiplier depending upon where on the photocathode the photon strikes. In Fig. 1.4, a limiter was used to trim the white dots to nearly the same size.

Figure 1.6 shows clearly that the boundary between the discernable and the nondiscernable parts of Fig. 1.5 lies approximately along a 45° diagonal. Further, the boundary moves one step to the right, toward discs that decrease by a factor of 2 in diameter, for each factor of 4 increase in light intensity, as is to be expected from Eq.(1.2). The series of pictures in Fig. 1.6 was used in the early publication[R-1] to estimate a value for k, the threshold

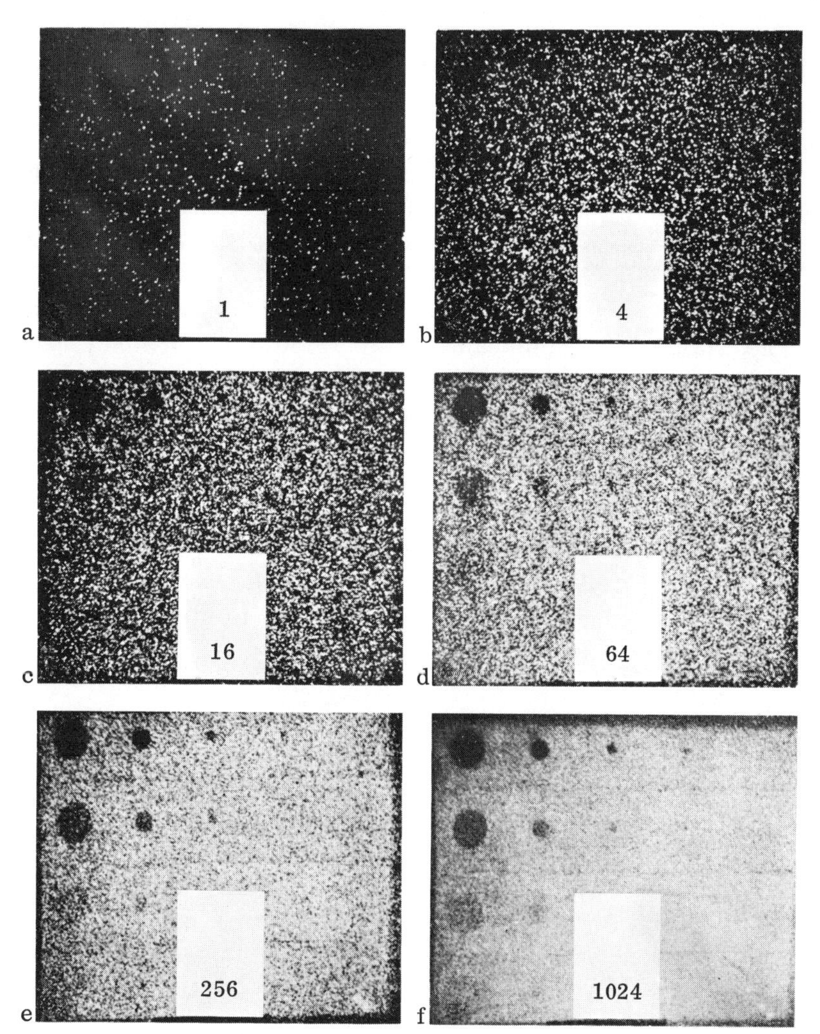

Fig. 1.6. Reproduction of Fig. 1.5 using a light-spot scanning arrangement in which the trace of single photons was made visible. The relative numbers of photons are indicated on each photograph.

signal-to-noise ratio, lying between 4 and 5. As we noted earlier (see Table 1.1), the value of k should increase from 3 to 5 in going from very low densities of photons for which the number of picture elements is less than 10^3 to high densities of photons for which the number of picture elements exceeds 10^5.

1.5. Resolution, Signal-to-Noise Ratio, and Test Patterns

The term "picture element" has been used here as meaning the smallest area of a spot of a given contrast that can be resolved. The shape of the spot is not critical; it can be round, square, or even rectangular. The area is of primary significance in determining the signal-to-noise ratio which is based on the average number of photons falling on that area.

It is clear by inspection of Fig. 1.6 that the term picture element has a certain elastic significance. If we are looking for small black spots on a white surface, the number of discernable picture elements is larger than if we are looking for gray spots having a small contrast with the white surround. From Fig. 1.6 or from Eq. (1.2), the number of discernable picture elements is proportional to the square of their contrast. There are 10^4 more picture elements having a contrast of 1 (black spots) than those having a contrast of 0.0l, or 1%. A lack of recognition of this relationship was responsible for many years for an inflated estimate of the resolving capability of photographic film.

The quoted values for film resolution were (and also, frequently, still are) based on the smallest spacing of a set of *black and white bars* that could be resolved. It was not unusual, for example, to use a given film with a rating of say 2000 lines per picture and find that not even details having the dimensions of 400-line resolution could be resolved. The primary reason, of course was that the detail to be resolved was low contrast, and not black and white. A second reason is that the detail was in the form of a single picture element and not in the form of a set of bars. The use of bars, rather than the single spots of Fig. 1.5, leads to a gross overstatement of the resolving properties of a system. The bars yield a higher resolution because the estimate of signal-to-noise ratio tends to be based on the total area of the bars rather than on a single elemental area whose diameter is the width of one bar. In Fig. 1.4, for example, the smallest black spots are not visible in Fig. 1.4c. At the same time, the presence of a long bar, whose width is slightly less than that of one of the smallest black dots, is readily visible.

There was a period during the early history of television

when members of the motion picture industry asked for television channel widths of the order of a hundred megacycles in order to transmit their films.[(K-1)] This was based on a black and white bar-pattern resolution rating for their film of 2000 lines. A casual inspection of current television pictures shows that viewing this same 2000-line film through the present 500-line (5 megacycle channel width) television standards frequently *degrades* the picture quality as compared with live studio pictures. That is, the signal-to-noise ratio of the film at 500 lines is frequently less than that which emerges from the studio camera transmitting live scenes.[(H-1)]

The misleading effect of using a test pattern of bars to define the resolution of a system is clearly brought out in Fig. 1.7. This is a series of pictures of ten pairs of black and white bars taken by Coltman.[(C-1)] The number under each picture gives the relative illumination of the test pattern. The illumination was sufficiently small (and the gain of the system sufficiently high) that each white dot represents a photon. It is clear that even in the first picture of the series, the one which has the lowest illumination, the presence of the bars can be detected, for example, by viewing the pattern somewhat edgewise, first from the side and then from the bottom. In this picture the density of photons is so low that if we defined a picture element as a square element whose length of side is the width of one bar, the average number of photons in this element would be significantly less than unity and in the neighborhood of $\frac{1}{4}$. The signal-to-noise ratio of such an element would then be $(\frac{1}{4})^{1/2}$, or 0.5. This is a misleading use of the term signal-to-noise ratio since it applies to an element which is small compared to what the eye is actually looking at or making use of in order to

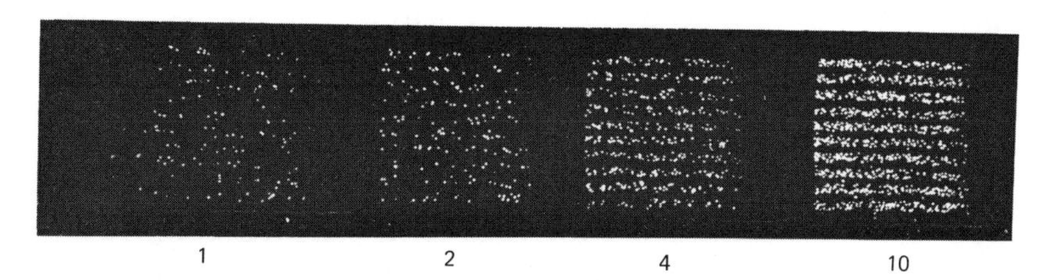

Fig. 1.7. Enhanced visibility of bar patterns as compared with dot patterns (Coltman).[(C-1)]

determine the presence of the bar patterns. The eye is making use of a large part of the pattern in order to achieve a signal-to-noise ratio well above unity. Single isolated elements whose dimension is the width of a bar would, of course, be completely undetectable in the first and also in the second picture of this series.

Consider also the first of the series of pictures in Fig. 1.6. This picture was recorded from the kinescope of a 500-line television system. If we were to estimate the signal-to-noise ratio of an element whose diameter is one television line width, the value, based on the density of photons, would be far less than unity. Since there are some few thousand photons in the entire picture, and a few hundred thousand television picture elements, the average number of photons per picture element is 10^{-2} and the signal-to-noise ratio referred to these elements is 10^{-1}. It is clearly evident from Fig. 1.6 that elements of such small dimension and small signal-to-noise ratio are completely beyond the range of visibility. It is only by going to the area of the largest black spot that we attain a threshold visibility and a threshold signal-to-noise ratio of about 4.

In brief, the signal-to-noise ratio of what the eye is able to detect must be well in excess of unity. Signal-to-noise ratios less than unity, as reported by Coltman[C-2] or by Morgan[M-1] for the viewing of bar patterns, may have a certain utility as reference numbers, but they do not define the signal-to-noise ratio of what the eye actually apprehends.

Parenthetically, there is some ambiguity about the meaning of signal-to-noise ratios less than unity. For example, the signal-to-noise ratio of 0.1 cited above was associated with an average density of 0.01 photons per picture element. Hence, on the average, a viewer would see zero photons in the picture element 99 times out of 100 observations. A single observation does not, in general, give any information about the magnitude of signal-to-noise ratios less than unity.

Another frequent error arises in motion picture or television practice when the signal-to-noise ratio of a picture element in a *single* frame is associated with threshold visibility observations on the moving film. If the storage time of the eye were equal to the time for which one frame is viewed in motion pictures, the above association would be valid. As it is, the storage time of the eye is about 0.2 sec and the time for one frame (in a television system) is 0.03 sec. Hence, the eye is in effect looking at the superposition of some 7 successive frames and achieving a signal-to-noise ratio which is larger than that of a single frame by the factor $\sqrt{7}$. Anyone who has looked at a single frame of a motion picture is immediately aware that it is noisier than the visual impression gained from the moving film in normal projection.

1.6. An Absolute Scale of Performance

Using Eq. (1.2), an absolute scale of performance can be plotted against which the performance of any actual picture-seeing device or system can be measured. We choose a fixed value of 5 for k, the threshold signal-to-noise ratio, with the understanding that in the region of low light levels its value should be somewhat lower. There are no reliable measurements on the variation of k with light level. Further we compute Eq. (1.2) for 1 cm^2 of image surface. Hence Eq. (1.2) can be written

$$\text{Number of photons/cm}^2 = n = \frac{25}{d^2C^2}$$

or

$$\text{Number of resolvable lines/cm} = \frac{1}{d} = \frac{Cn^{1/2}}{5} \tag{1.3}$$

The term "number of resolvable lines/cm" is used here to measure the diameter d of the smallest resolvable isolated spot having a contrast C. It does not refer to the customary operation of viewing bar patterns with spacing d. The latter, as we have pointed out, are intrinsically more visible than single isolated spots. Figure 1.8 is a plot of number of lines/cm *versus* C with the photon density n as a parameter.

If we know, for example, that the image surface of some visual system has received an exposure of 10^{10} photons/cm^2, then, according to Fig. 1.8, we should be able to resolve black and white elements ($C = 1$) whose diameter is greater than $\frac{1}{2} \times 10^{-4}$ cm. At the same time, we should only be able to resolve elements, having a contrast of 0.01, whose diameter is greater than $\frac{1}{2} \times 10^{-2}$

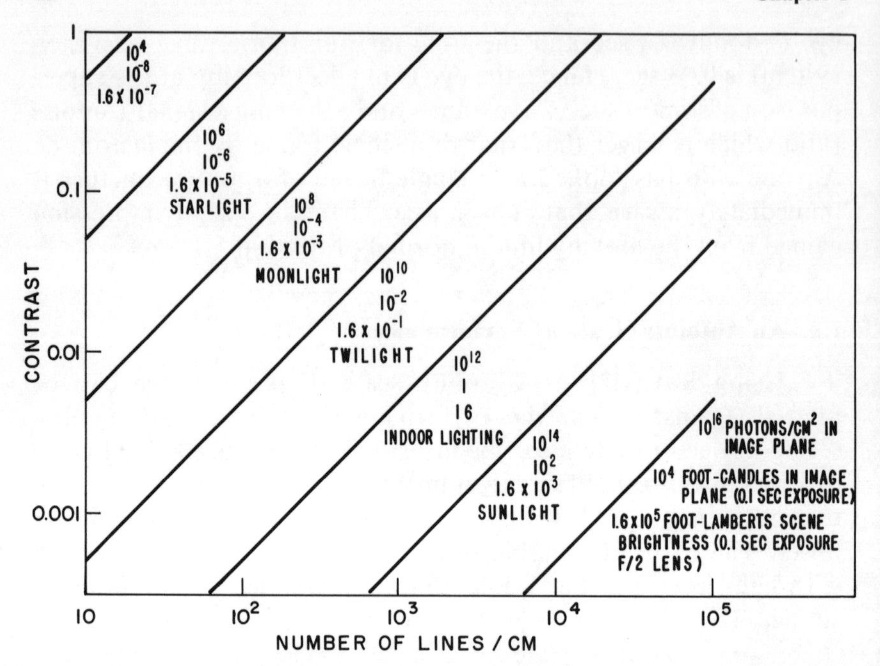

Fig. 1.8. Universal plot of the limiting size and contrast of single elements that can be transmitted at various photon densities.

cm. All of this would be true if, indeed, the visual system counted every incident photon. Under these conditions its quantum yield would be unity (a quantum *yield* of unity equals a quantum *efficiency* of 100%).

Now suppose that, in fact, the smallest black and white element that can be resolved by the visual system with 10^{10} photons/cm^2 on its image surface is only $\frac{1}{2} \times 10^{-3}$ cm rather than $\frac{1}{2} \times 10^{-4}$ cm. Then, according to Fig. 1.8, the effective performance is that of a system having a quantum yield of 0.01, or a quantum efficiency* of 1%.

The parameters of the curves in Fig. 1.8 are given primarily in photons/cm^2 in the image plane. For convenience, the photons/

* Quantum efficiency, here, has the same meaning as the phrase DQE (detective quantum efficiency) which was introduced by R. Clark Jones and which is used extensively in the literature.

cm^2 are converted also to foot-candles illumination in the image plane for an exposure time of 0.1 sec.* The conversion factor used is 10^{16} photons/sec per lumen of white light. A third equivalent shown on the curves is the corresponding scene illumination in foot-lamberts when an *f*/2 lens is used and an exposure time of 0.1 sec. Finally, the locations of various representative light levels are indicated from starlight to bright sunlight.

Figure 1.8 shows the performance of ideal noise-limited visual systems. It is likely that the resolving power of actual systems for small black test elements will be limited by lens errors, by diffraction, or by structure in the image plane. Similarly, the ability of actual visual systems to portray small contrasts in large areas may be limited by various sources of system noise such as nonuniformities in the recording medium. The result is that the performance curve for an actual system may not lie along a 45° line, but may be bowed such that the high-resolution and low-contrast ends show a lower performance than some intermediate part of the curve. Figure 1.6 shows evidence of this bowed type of cutoff. It will appear again in the peformance curves for human vision.

1.7. Geometric *versus* Noise Limitations to Performance

The preceding remarks are presented schematically in Fig. 1.9. The solid lines labeled "signal" give the amplitude response of an imaging system as a function of the number of lines/cm in a test pattern. These curves are a measure of the geometric limitations of the system such as diffraction, lens aberrations, and the finite size of the elements of the imaging surface. The curve labeled "rms noise" is the noise current that would be observed if the image were scanned by a series of apertures, each corresponding to a certain number of lines/cm. (The finer apertures give larger noise currents, increasing as the number of lines/cm, or the reciprocal width of the aperture. The reason is that even though the rms fluctuation within the aperture decreases as the aperture dimension, the linear velocity at which the aperture scans the image must increase as the reciprocal area, or as the square of its

* We assume here a lens transmission and a scene reflectivity of 100%.

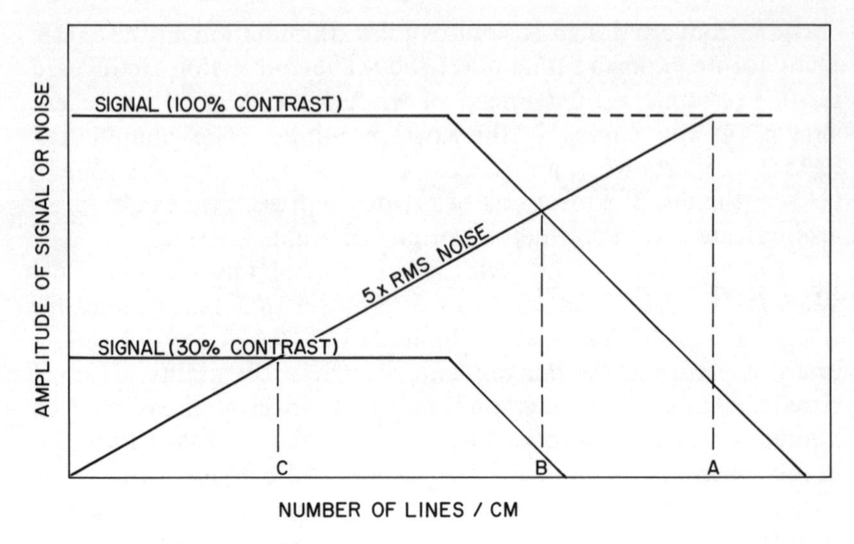

Fig. 1.9. Schematic comparison of geometric and noise limitations to resolution.

reciprocal dimension, in order to cover the entire image in a fixed time. The product of these two effects then yields a noise current increasing as the reciprocal of the aperture width. By the same arguments, the signal current is independent of aperture size.) The noise curve is plotted as 5 × rms noise in order that its intersection with the signal curve will yield directly the smallest resolvable elements.

If there were no geometric limitations to the resolution of the system, the smallest resolvable elements would lie at point *A*. Actually, the drop in the geometric resolution causes the cutoff to lie at a lower line number, at point *B*. This is true for picture elements having 100% contrast, that is, black and white. If we turn our attention to low-contrast elements [curve marked "signal (30% contrast)"] we find that the cutoff lies at point *C* where the geometric response is still unimpaired. The same is true for low-light scenes in general.

The point we wish to make is that even when geometric limitations on the resolution of a system begin to play a role, they affect selectively only the high-contrast parts of the picture. The visibility and signal-to-noise ratio of the low-contrast parts are likely still to be unaffected; and most of the information in an average picture is of low contrast.

Our emphasis throughout this monograph is on the limitations imposed by the finite number of photons rather than on the less fundamental limitations imposed by the finite geometric response of the system.

1.8. Beyond the Visible Spectrum

The arguments outlined in this chapter have been concerned primarily with radiation in the visible range of wavelengths (0.4–0.7 μm). Since the arguments have been couched in terms of numbers of photons, they apply equally well to any system that can detect ultraviolet radiation, x-rays, or gamma rays. Sturm and Morgan,[S-1] for example, have given an excellent discussion of the information transmitted by a finite number of x-ray photons. In the case of visible and higher energy radiations, one can extend the arguments down to almost arbitrarily low densities of photons since the thermal densities of these photons are vanishingly small.

One can also apply the arguments to the range of infrared radiation. Here, however, one must contend at ordinary temperatures with a significant density of thermally generated photons. It is as if the surround were never dark. Indeed, the ambient density of photons whose wavelength is in the neighborhood of 10 μm is comparable with that of bright sunlight, namely, about 10^{18} photons/cm^2-sec incident on or emitted from a surface. At 3 μm, the photon density is of the order of that for room light, or about 10 foot-lambert, and at 1 μm the density is equivalent to a visible ambient density of about 10^{-11} foot-lambert, that is, far below the absolute threshold for vision.

In general, the photon flux emitted by a blackbody at temperature T is

$$\frac{\Delta\nu}{\lambda^2}\exp\left(-\frac{h\nu}{kT}\right)\text{photons}\cdot\text{cm}^{-2}\cdot\text{sec}^{-1}\,\text{sr}^{-1}$$

where $\Delta\nu$ is the range of optical frequencies in the neighborhood of the wavelength λ.

The visibility of objects in the infrared region is a complex function of the artificial illumination used, their self-luminous flux, their emissivities, and their temperature differences. We mention here only that the contrast of objects viewed by their own radiation and having the same emissivities is

$$\frac{h\nu}{kT}\frac{\Delta T}{T} \times 100\%$$

where ΔT is the temperature difference between an object and its surround. A temperature difference of 1 deg (centigrade) yields a contrast of about 10% at wavelengths of 1 μm and a contrast of about 1% at 10 μm.

1.9. Summary

The information content of a finite amount of light is limited by the finite number of photons, by the random character of their distribution, and by the need to avoid false alarms.

The signal-to-noise ratio of a test element in an image is defined as the ratio of the average number of photons in the element (or difference in average numbers between it and the surround) to the rms deviation from the average. For an average number n, the signal-to-noise ratio is $n^{1/2}$.

The threshold signal-to-noise ratio k is the ratio of signal to noise required to avoid false alarms. Its value is normally about 5 and may be as low as 3 under extreme low-light conditions.

The characteristic for an ideal photon-noise-limited system is

$$nd^2C^2 = k^2$$

where n is the number of photons/cm^2, d is the diameter of test element, C $(= \Delta B/B)$ is the contrast of the test element with the surround, and k $(= 5)$ is the threshold signal-to-noise ratio.

The signal-to-noise ratio of a system has meaning only when the size of the test element has been specified.

The resolution of a system has meaning only when the contrast of test element has been specified.

Geometric limitations on resolution affect the high-contrast elements more than the low-contrast elements.

Bar patterns are more visible than an isolated spot whose diameter is equal to the width of the bar.

1.10. References

C-1. J. W. Coltman, Scintillation limitations to resolving power in imaging systems. *J. Opt. Soc. Am.* **44,** 234 (1954).

C-2. J. W. Coltman and A. E. Anderson, Noise limitations of resolving power in electronic imaging. *Proc. IRE* **48,** 858–865 (1960).

H-1. L. Hayen and R. Verbrugghe, A comparison of the signal-to-noise ratio and sensitivity of film and plumbicon camera, *J. Soc. Motion Picture Television Engrs.* **81,** 184 (1972). The authors find that the signal-to-noise ratio of a studio television picture exceeds that of 16-mm, color reversal film, ASA-125. Many of the 16-mm film clips used on television are significantly noisier than the color film used by these authors.

K-1. I. J. Kaar, The road ahead for television, *J. Soc. Motion Picture Engrs.* **32,** 18 (1939).

M-1. R. H. Morgan, Threshold visual perception and its relationship to photon fluctuation and sine-wave response, *Am. J. Roentgenol., Radium Therapy Nucl. Med.* **93**, 982–996 (1965).

R-1. A. Rose, The sensitivity of the human eye on an absolute scale, *J. Opt. Soc. Am.* **38,** 196 (1948).

S-1. R. E. Sturm and R. H. Morgan, Screen intensification systems and their limitations, *Am. J. Roentgenol. Radium Therapy* **62,** 617 (1949).

General

R. Clark Jones, "Quantum Efficiency of Detectors for Visible Infrared Radiation," in *Advances in Electronics and Electron Physics,* Vol. 11, pp. 87–183 (1959), Academic Press, New York.

A. Rose, "Television Camera Tubes and the Problem of Vision," in *Advances in Electronics and Electron Physics,* Vol. 1, pp. 131–166 (1948), Academic Press, New York.

A. Rose, "Quantum Effects in Human Vision," in *Advances in Biological and Medical Physics,* Vol. 5, pp. 211–242 (1957), Academic Press, New York.

O. H. Schade, The resolving-power functions and quantum processes of television cameras, *RCA Rev* **28**, 460–535 (1967).

R. E. Sturm and R. H. Morgan, Screen intensification systems and their limitations, *Am. J. Roentgenol. Radium Therapy* **62**, 617 (1949).

CHAPTER 2

HUMAN VISION

2.1. Introduction

The human visual system has attained a remarkable level of sophistication. The scientists engaged in matching its performance by some electronic or chemical system can only marvel at its sensitivity, its compactness, its long life, its high degree of reproducibility, and its elegant adaptation to the needs of the human system. It is true that the attempts at man-made systems date back only a scant hundred years, while the human visual system has been millions of years in the making. But the human visual system had to emerge from some cosmic scrambling of the elements, repeated and repeated and repeated until by chance the happy combination fell into place. While there are few who would question the blind, probabilistic, evolutionary origin of the human species, there are none who can delineate the steps. The shortfalls have long since vanished without a trace.

In the scheme of evolution, vision has an almost unique role. One can conceive, for example, that further evolutionary developments might lead to a larger brain capacity, a more involved nervous system, or a variety of enhancements of current functions. It is *not* conceivable that the sensitivity of the visual process can be significantly enhanced. The visual process is at an absolute terminal point in the evolutionary chain. To the extent that the visual process now succeeds in counting each absorbed photon, there is little possibility, outside of increasing the absorption, of a further increase in sensitivity. The laws of quantum physics impose a firm limit which has been closely approximated by our existing visual apparatus.

We made the proviso that vision has an *almost* unique role because there is evidence that some of the other sensory processes have also evolved against an absolute limit. The ability of insects (moths), if not other species, to detect single molecules is evidence that the sense of odor has in some instances reached a quantum limit. Similarly, our sense of hearing is limited in the extreme by the presence of an ambient thermal noise.

The high sensitivity of the visual process is not confined to the human species. The lower forms of animals and the nocturnal birds show clear evidence of comparable performance. Presumably, the species of fish that inhabit the deeper and darker reaches of the oceans must also exhaust what little information filters down through the straggling wisps of light. Finally, we can point to photosynthesis as evidence that the early forms of plant life already succeeded in making use of almost every incident photon, at least within a certain spectral range.

The primary aim of this chapter is to present the data on human vision in such a form as to demonstrate the high quantum efficiency of the eye over a wide range of light intensities. In order to convert raw visual data into photons/cm^2 on the retina, it is necessary to know the optical parameters of the human eye. These are reviewed in the next section.

2.2. Optical Parameters

An outline of the dimensions of the human eye is shown in Fig. 2.1. The lens has a pupil opening that varies from 2 mm at high ambient light levels to about 8 mm near the threshold of vision. The pupil opening adjusts within tenths of a second to the ambient light level. The focal length of the lens is 16 mm. This means that the speed of the optical system varies from $f/2$ at low light levels to $f/8$ at high light levels. A representative curve of pupil opening *versus* light level is shown in Fig. 2.2.[R-1]

The light-sensitive layer, called the retina, is composed of discrete light-sensitive cells, the rods and cones, spaced about 2 μm apart. There are some 10^8 of these elements in the total retinal area

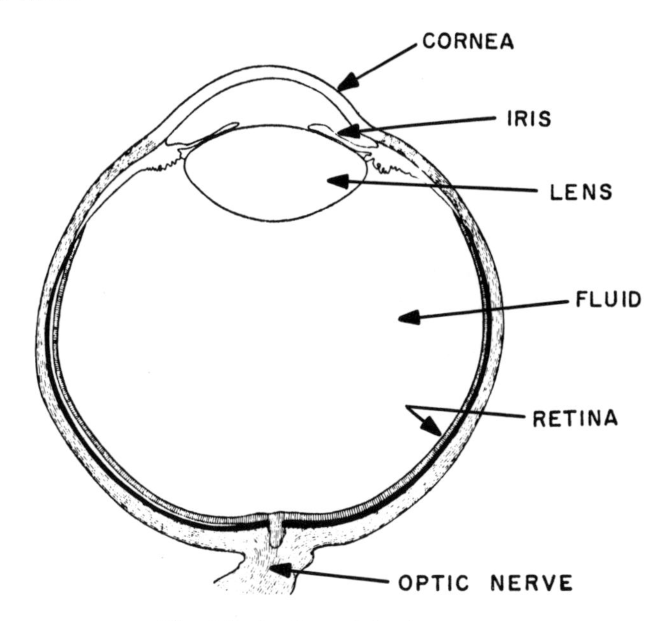

Fig. 2.1. Outline of the human eye.

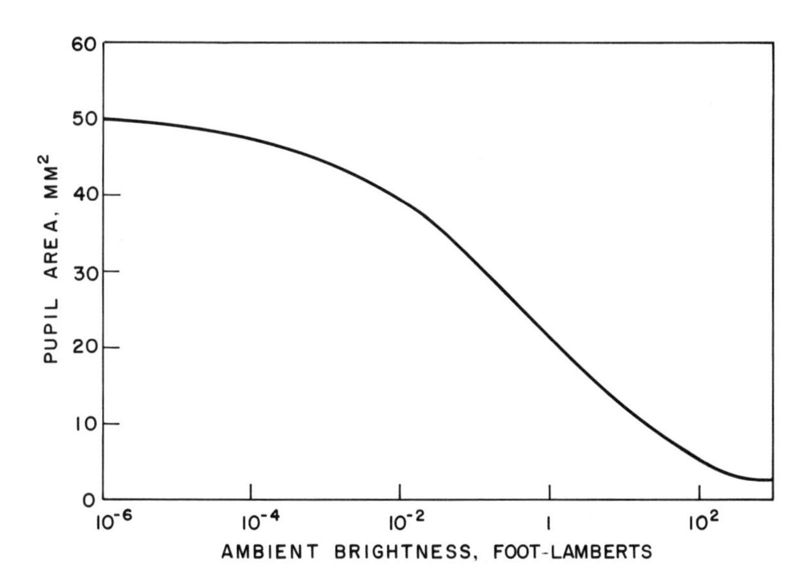

Fig. 2.2. Plot of pupil opening as a function of ambient brightness.

of about 10 cm^2. The cones, located predominantly in the central or foveal area subtending about one degree, are active at medium and high light levels and transmit color sensations. The rods, occupying most of the retinal area, are active down to the lowest light levels and are devoid of color sensation. The cones set the limit of resolution at high light levels to about 1 or 2 minutes of arc and closely match the diffraction disc appropriate to a 2-mm lens opening. There is operational and anatomical evidence that the rods are connected in larger and larger groups up to several thousand elements as one departs from the center of the retina. Note that the light striking the retina passes through the layer of nerve fibers fanning out from the optic nerve to the retinal cells.

The space between the lens and the retina is filled with an aqueous medium whose index of refraction is 1.5. It is estimated in the literature that half the light incident on the eye reaches the retina. The rest is reflected or absorbed.

The physical storage time of the eye is between 0.1 and 0.2 sec, and is probably closer to the latter. The physical storage time is the equivalent of exposure time in a photographic camera. There is, perhaps, at most a factor of 2 increase in storage time in going from high light levels to the threshold of vision. The eye obeys a reciprocity law so that it senses only the product of light intensity and time for times less than 0.1 or 0.2 sec.

2.3. Performance Data

There has been a continuous outpouring of data on human vision over the past hundred years. Some of the more recent and complete measurements of the ability of the eye to detect a single spot of varying size and contrast under a wide range of illumination have been published by H. R. Blackwell.[B-3] Figure 2.3 is a plot of Blackwell's data for the range of scene illuminations from 10^{-6} to 10^2 foot-lamberts, range of contrasts from 1% to 100%, and range of angular resolution from 3 to 100 minutes of arc. We have omitted data for contrasts less than 1% or angular resolution less than 3 minutes of arc since visual performance in these ranges is clearly not limited by noise considerations, but by other constraints which set the absolute limit of contrast discrimination at $\frac{1}{2}$% and angular

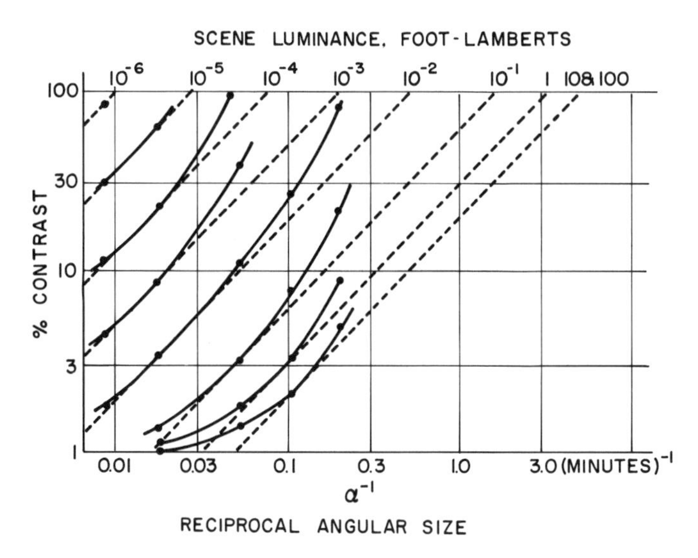

Fig. 2.3. Visual performance data as measured by Blackwell.[B-3]

resolution at 1–2 minutes of arc. In the latter case, the finite size of rods and cones set the geometric limit to resolution.

Figure 2.4 is a plot of similar and earlier data by Conner and Ganoung (1935)[C-2] and Cobb and Moss (1928).[C-1] A comparison of Figs. 2.3 and 2.4 shows substantial agreement. One notable difference is that Blackwell's data show no improvement in going from 10 to 100 foot-lamberts, whereas the data of Cobb and Moss do.

In both Figs. 2.3 and 2.4, 45° lines represent the performances to be expected if the performance were noise limited as shown by Eq. (1.2). The data in Fig. 2.4 follow the noise-limited 45° slopes moderately well. The data in Fig. 2.3 are somewhat bowed and only tangent to the 45° slopes in a finite range. The departures in Fig. 2.3 from the 45° slopes can be ascribed to the admixture of limitations other than photon noise.

2.4. Quantum Efficiency of Human Vision

The data in Figs. 2.3 and 2.4 need to be converted into photons/cm^2 at the retina in order to make an estimate of the quantum

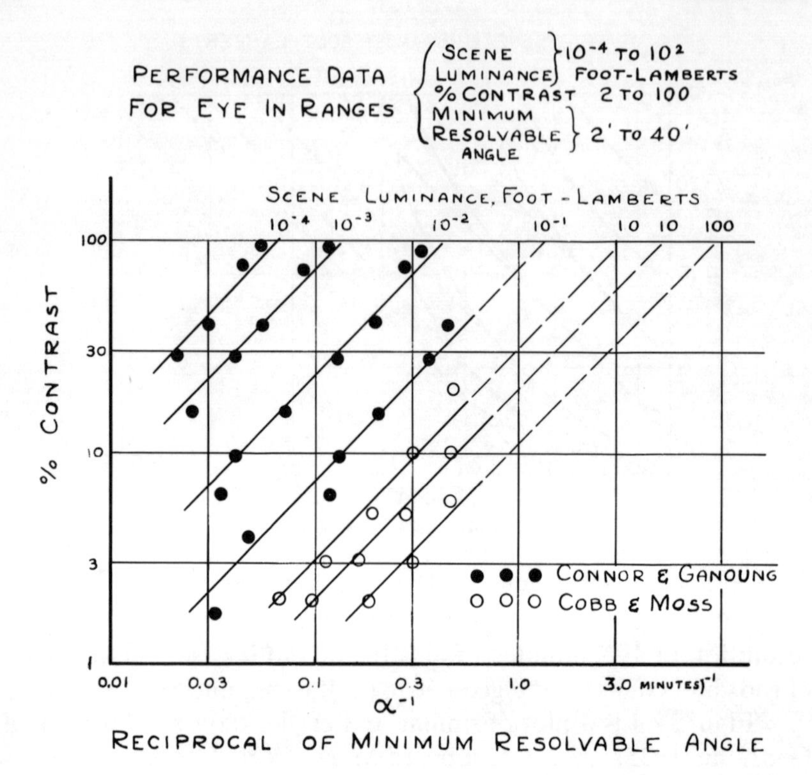

Fig. 2.4. Visual performance data as measured by Cobb and Moss[(C-1)] and by Connor and Ganoung.[(C-2)]

efficiency of the eye. To do so we have assumed a storage time of 0.2 sec, a lens transmission of 0.5, and a range of pupil openings given by Reeve's data in Fig. 2.2. Once having made the conversion, the photon density at the retina is inserted into Eq. (1.3) in the form

$$C^2 d^2 \theta n = k^2 = 25$$

where θ is the quantum yield of the eye (quantum efficiency $\equiv 100 \times \theta\,\%$). The value of k, the threshold signal-to-noise ratio, is taken to be 5.

Figure 2.5 shows the quantum efficiency of the eye, computed from Blackwell's data, as a function of ambient brightness. The most striking aspect of these results is the relatively small variation of quantum efficiency in the span of 8 orders of magnitude of light

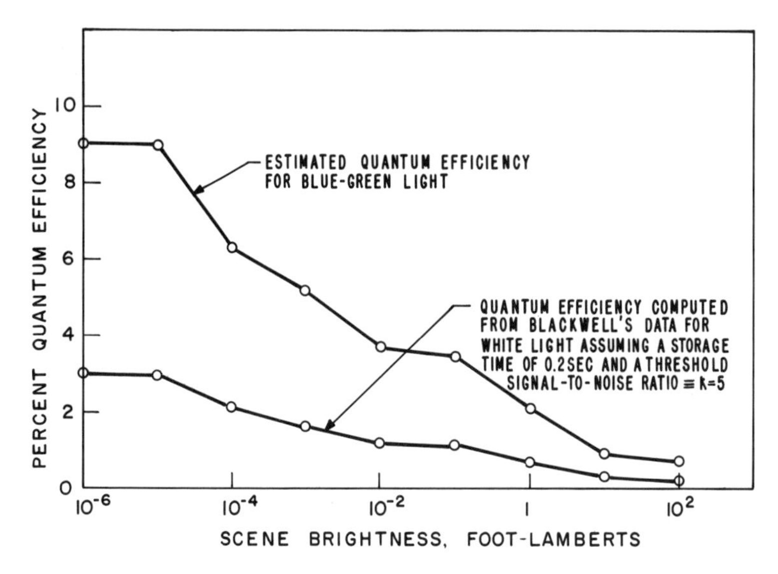

Fig. 2.5. Quantum efficiency of the eye as a function of ambient brightness. The values for gree-blue light at the peak of visual sensitivity are taken to be three times larger than for white light. The quantum efficiencies at the retina should be about twice the values plotted above. The sum of the photons incident on both eyes was used in computing the quantum efficiency.

intensity. The quantum efficiency starts at 3% at extreme low lights near absolute threshold (approximately 10^{-7} foot-lambert) and declines slowly to about $\frac{1}{2}$% at 100 foot-lamberts. It is true, on the one hand, that this is a tenfold variation in quantum efficiency. On the other hand, it must be compared with a 1000- or 10,000-fold variation in quantum efficiency which was invoked in the earlier literature to account for the phenomenon of dark adaptation. We will return to this point in more detail below. In the meantime, it is well to recognize that even this tenfold variation may considerably overstate the facts. In deducing the tenfold variation we have assumed a constant exposure time and a constant k factor. There is some evidence that the exposure time may be a factor of 2 larger at low lights than at high lights. If so, the variation of quantum efficiency would then be reduced to only fivefold. It is further likely that the k factor is smaller at low lights than at high lights. This variation in k (or, rather, k^2) could easily introduce another factor of 2 which would bring the overall variation in quantum efficiency down to a mere factor of 2 in a range of 10^8 in light intensity.

The second major aspect of the quantum efficiencies in Fig. 2.5 is their relatively large magnitude. There have been estimates in the literature[R-7] that the sensitive material (rhodopsin) in the retina absorbs only 10% of the incident light. If so, the quantum efficiency (for white light) referred to *absorbed light* would then be in the order of 60% at low lights. This would leave scant room for improvements in the photon counting mechanism itself. It is, however, puzzling to understand what evolutionary purpose was served in absorbing only 10% of the incident light. It is possible that the limited availability of otherwise compatible biological materials is the source of this curiously low absorption.

Some of the drop in quantum efficiency at high lights can be ascribed to the demands of a color-sensitive system. If, as recent evidence has shown,[D-1] there are 3 sets of cones with different spectral responses, the receptive area for light of a given wavelength would be reduced at high light levels by as much as a factor of $\frac{1}{2}$.

The quantum efficiencies in Fig. 2.5 (lower curve) are based on white light. It is known that the visual response to green light is some three times higher than the response to the same total number of "white" photons—that is to photons distributed across the visible spectrum. The use of green light (or greenish-blue at low light) should have increased the quantum efficiencies as shown in Fig. 2.5 by a factor of 3. In this event the quantum efficiency at low lights would be some 10% and would force the estimate of the fraction of light absorbed by the retina to be at least 20%, rather than 10%, of that incident on the retina.

We would like to re-emphasize that the quantum efficiencies of Fig. 2.5 are dependent on the choice of the parameters 0.2 sec for the storage time of the eye and $k = 5$ for the threshold signal-to-noise ratio. These values are subject to some uncertainties, particularly as they apply to Blackwell's data. It is possible that improved values for the parameters can lead to higher values of quantum efficiency. For example, the assumption of a storage time of 0.1 sec would double the quantum efficiencies in Fig. 2.5. However, it is doubtful that the effort spent on refining these parameters can compare in value with the effort spent on an improved experimental format for measuring quantum efficiency which is independent of the parameters.

2.5. A Preferred Method for Measuring Quantum Efficiency

There is now available a remarkably simple arrangement for measuring the quantum efficiency of the eye with some confidence. The recently developed silicon intensifier television camera tubes (see Chapter 3) are able to transmit pictures at low levels of illumination that are clearly photon-noise limited, that is, limited by the noise of that fraction of incident photons that give rise to photoelectrons at the photocathode. The significant fact is that the camera tube can present an ideally photon-noise-limited picture appropriate to the reliably measurable quantum efficiency of the photocathode. The procedure, then, is to have an observer and a television camera view the same low-light scene at the same viewing distance. The lens opening on the camera is matched to the pupil opening of the human observer. The observer then compares what he sees looking directly at the low-light scene with what he sees on the kinescope of the television system. If the information is the same, the observer has the same quantum efficiency as that measured on the photocathode of the camera tube. If the observer sees more or less than the camera, the lens opening on the camera can be adjusted until there is a match, and the quantum efficiency of the observer calculated from the relative lens openings.

The profound virtue of this side by side comparison is that it does not depend upon the visual exposure time or upon a proper choice of threshold signal-to-noise ratio. These parameters, whatever their proper values, remain substantially the same for the observer's examination of the original scene as for the scene presented on the television screen. Hence, they cancel out in the comparison. Furthermore, the significant contribution that memory can make to the effective exposure time can be expected to be substantially the same for the two observations.

We emphasize this tool because it should now be readily available to experimenters skilled in the visual process. Comparable arrangements have been used both by the author and by others to make informal estimates of the quantum efficiency at low lights. The author[R-3] used a light-spot scanner (Fig. 2.6); Dr. J. E. Ruedy[R-6]

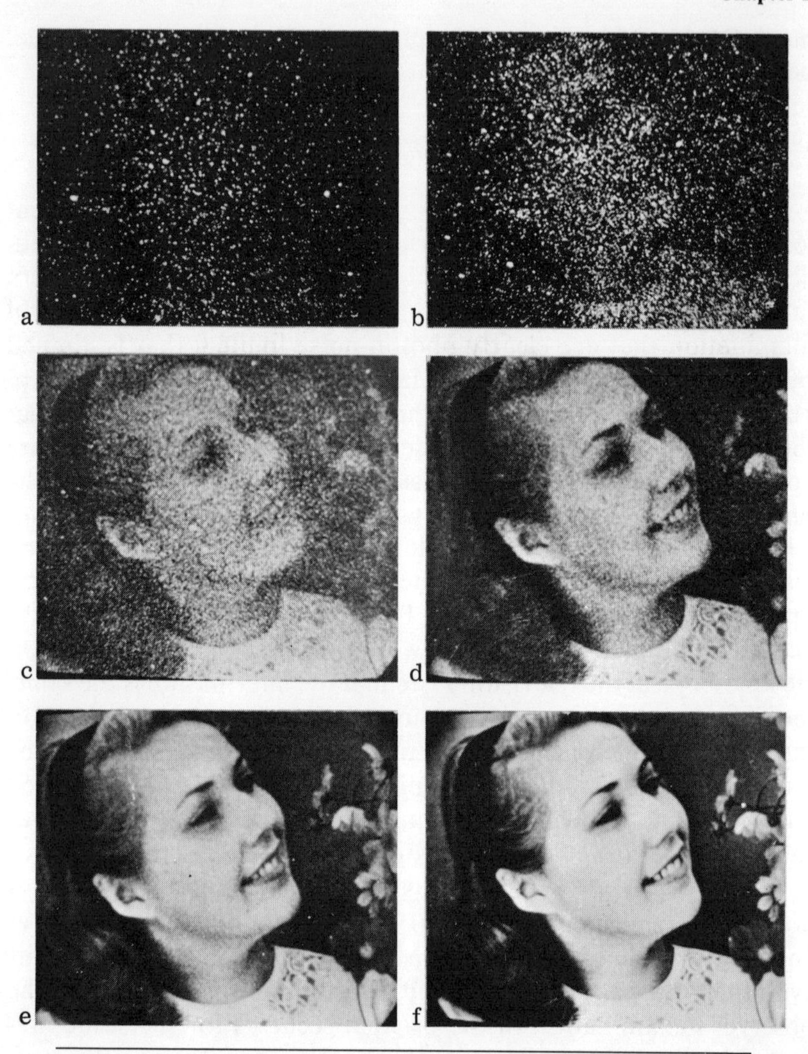

Picture	Number of photons	High-light brightness, foot-lamberts
a	3×10^3	10^{-6}
b	1.2×10^4	4×10^{-6}
c	9.3×10^4	3×10^{-5}
d	7.6×10^5	2.5×10^{-4}
e	3.6×10^6	1.2×10^{-3}
f	2.8×10^7	9.5×10^{-3}

Fig. 2.6. Series of pictures used in evaluating the quantum efficiency of the eye.(R-4)

used an intensifier image orthicon, and Dr. G. T. Reynolds[R-2] used a multistage image intensifier. All of these arrangements delivered photon-noise-limited pictures and all of the informal estimates of quantum efficiency were in the neighborhood of 10% at low levels of illumination.

The series of pictures in Fig. 2.6 shows the maximum amount of information that can be conveyed by various known numbers of photons. Each photon is recorded as a discrete visible speck. Only the statistical fluctuations, inherent in the recording of a photon stream, limit the perceivable information. The numbers of photons N refer to the total number of photons in the picture if the picture were uniformly at its high-light brightness.

The luminance values assigned to the pictures were computed with the assumption that the eye makes full use of one out of every ten incident photons. The other parameters used to compute the luminances are a storage time of 0.2 sec and a pupil diameter of about 6 mm. In other words, if one replaced the subject by a white card having the luminance indicated, computed the number of photons entering the eye in 0.2 sec, and divided this number by 10, the result would be the number of photons N associated with that value of luminance. Accordingly, this series of pictures shows the maximum amount of information that an actual observer can perceive at each luminance if the quantum efficiency of his visual process is 10% and his viewing distance from the life-size subject is 4 ft.

2.6. A Comparison of Estimates of Quantum Efficiency

It has been known for a century or more that at the absolute threshold of vision, a burst of some 100 photons incident on the eye from a small source is just visible. This put a lower limit on the quantum efficiency of about 1%. There followed, then, a series of experiments by several groups of investigators to determine how many of the 100 photons the eye actually used. If, for example, the eye used all 100 photons, the transition from not seeing to seeing would be quite sharp as the incident photon flux was increased from below to above 100. If the eye used only a few photons, this transition would be characteristically diffuse owing to the stochastic nature of photon emission. The sharpness of the transition, then, is a measure of the number of photons used, and consequently of the quantum efficiency of the eye.

The concept of the experiment had a certain simplicity and elegance. Its execution led, unfortunately, to a wide range of estimates for the number of photons actually used by the eye in forming its threshold sensation.[B-4, H-1, V-1] The spread extended from 2 to 50 photons and left the question of quantum efficiency correspondingly diffuse. It is perhaps not surprising to an electronics engineer or physicist that these observations had such varied results. The measurements were taken near the absolute threshold of vision where the noise of the photon stream could easily be contaminated by extraneous sources of noise within the eye itself. For example, if one applied the same type of measurement to a photomultiplier tube, the estimates of quantum efficiency would show a similar divergence owing to the interference by noise from thermionic emission from the photocathode or by sporadic electrical breakdown between electrodes. All of this would be true if the measurements were made near the absolute threshold. If, on the other hand, a measurement of signal-to-noise ratio is made at a light level, well above the threshold, where the photon noise dominated the extraneous sources of noise, the procedure will lead to a reliable value for quantum efficiency. It is for this reason that the writer has stressed the greater reliability of measurements of visual quantum efficiency made at light levels well above the absolute threshold of vision.

R. Clark Jones[J-1] has analyzed the same data from which we deduced the quantum efficiency curve of Fig. 2.5. His quantum effiiciencies are generally about tenfold smaller than those of Fig. 2.5 and are based on a shorter integration time (≈ 0.1 sec) and much smaller value of k (1.2). Jones argued that since the observer had only to make one choice out of a possible eight positions for the test object, this value for k would yield 50% reliability. The argument is, of course, numerically correct. The central question is whether, indeed, the visual decisions of the observers were made in quite this fashion. In terms of Fig. 1.4a, a k value of 1.2 would mean that an observer could detect in which of eight possible areas an operator had removed one or two photons. Simple inspection of Fig. 1.4a makes this appear highly unlikely. In any event, it is just this type of question that emphasizes the need for a test procedure that would avoid the ambiguities as to the proper value for k or for the storage time. The side-by-side comparison of the performance of the human eye with a photon-noise-limited electronic device, described above, is such a procedure and merits further use by experimenters.

The early estimates of H. De Vries[(D-2)] of the visual quantum efficiencies were also based on a k value of unity and hence were considerably lower than those of Fig. 2.5. De Vries, however, was one of the first investigators to argue that photon noise could account for the observed resolution and halftone discrimination of the eye. He also, as has the writer, called attention to the fluctuating, grainy appearance of low-light scenes as being evidence of the granularity of light.

H. B. Barlow[(B-1)] avoided most of the ambiguity in the choice of k by making observations on two adjacent test spots of light. The object was to decide which spot was brighter as their relative intensities were varied. A statistical analysis of the results, on the assumption that photon noise limited the ability to discriminate brightness, led to values for visual quantum efficiences between 5 and 10% in the range of brightnesses up to 100 times above the absolute threshold of seeing. Barlow also quotes Baumgardt[(B-2)] and Hecht[(H-4)] as each arriving at quantum efficiencies of about 7% from an analysis of the frequency-of-seeing curve near the absolute threshold.

In summary, most investigators concur that the quantum efficiency of the human eye is between 5 and 10% at light intensities from the absolute threshold up to 100 times higher. This efficiency is for wavelengths near the peak of the visual response curve (blue-green) and refers to light incident on the cornea. If one assumes that only half the light reaches the retina, the efficiency at the retina would be 10–20%. Since the estimates of the fraction of light absorbed by the retina also lie in this range, one must conclude that the efficiency of the eye is essentially 100% when referred to the absorbed light. In brief, the eye is able to count each absorbed photon.

The data in Fig. 2.5 add the further highly significant fact that in the range from the absolute threshold to 100 foot-lamberts, a range of 10^8 in light intensity, the quantum efficiency decreases by at most a factor of 10. Further investigations may reveal that this factor is possibly no more than 2 or 3. We have, then, the impressive fact that the eye maintains a high level of quantum efficiency over a range of 10^8 in light intensity. We make use of this conclusion below in interpreting the phenomenon of dark adaptation and the appearance of visual noise.

2.7. Dark Adaptation

One of the most common and striking aspects of the visual process is that of dark adaptation. When one enters a dark theater from the glare of a city street, he finds himself virtually blind for some seconds or minutes. As time goes on, he begins to see more and more until after half an hour he is fully dark adapted. He is now able to see objects more than a thousandfold darker than those he could barely sense when he first entered the theater.

There is no doubt, on the face of the facts, that the "sensitivity" of the observer has increased more than a thousandfold in the course of dark adaptation. And, indeed, the scientific literature took its cue from these observations to look for a mechanism or a chemical model that would account for these wide swings in sensitivity. Hecht,[H-2] for example, put great emphasis on a reversible bleaching of the sensitive material, visual purple, in the retina. He argued that the visual purple was fully intact at low lights and thus had its maximum absorption. At high light levels, the visual purple was progressively bleached so that it absorbed less and less of the incident light. The long time required for dark adaptation was viewed as the time required to regenerate the high density of visual purple. In this way the eye regained its sensitivity.

In contrast to these arguments, the noise analysis[R-3] of the sensitivity of the eye showed that its intrinsic sensitivity could not change by more than a factor of 10 in going from a dark surround to a bright surround. The virtue of the noise analysis was that its conclusions were independent of the particular physical or chemical models for the visual process itself. The sensitivity was measured on an absolute scale whose only postulates were the quantum nature of light and the random character of photon distributions.

How then to account for the thousandfold or greater improvement in the ability to see during the course of dark adaptation? The form in which the answer must be couched was already familiar in such instruments as radio or television receivers. It was common, for example, to tune a radio or television dial from a strong to a weak station and find the sound at an almost inaudible level. The listener then reached for the volume control knob and brought the sound level of the weak station up to a comfortable level. What is significant in this series of operations is that the sensitivity of the radio receiver remained constant both in tuning from a strong to a weak station and in "turning up the volume." The sensitivity was already fixed by the electronic properties of the antenna and the first tube of the amplifier. The process of "turning up the volume" did not alter the sensitivity of the receiver, but only the level of presentation to the listener. To the extent that there was some time delay in resetting the volume, the entire operation of tuning from a strong to weak station was a complete parallel of the time-consuming process of visual dark adaptation.

The minutes required for dark adaptation are minutes during which the gain of the amplifier is being chemically raised to an appropriate presentation level. The intrinsic sensitivity of the eye remains substantially constant during the period of dark adaptation. There is little choice but to conclude that the visual process involves a high degree of amplification between the retina and the brain and that the gain of the amplifier is variable. At high lights the gain is small, at low lights it is large.

2.8. Automatic Gain Control

The necessity for an automatic gain control in the visual process was deduced in the previous section from the large changes in *apparent* sensitivity encountered during dark adaptation and the relative constancy of intrinsic sensitivity derived from the noise analysis of visual data.

The necessity for a high gain mechanism and for its variability could just as well have been deduced from other more direct evidence in the literature. The energy in a nerve pulse is known to be many orders of magnitude larger than that of the few photons required at the absolute threshold to trigger the nerve pulse. A correspondingly high gain mechanism is therefore needed immediately at the retina to generate nerve pulses. It has also been known from the early work of Hartline[(H-1)] in electrical recording of optic nerve pulses in the king crab that the rate of generating nerve pulses increases only logarithmically rather than linearly with increasing light intensity. This means that the gain is lower at high light intensities than at low light intensities.

While an accurate value for the energy in a nerve pulse is not known, an approximate value can be estimated on the assumption that the stored energy of the pulse is 0.1 volt across 10^{-9} farad, the capacitance across 1 cm of outer sheath of nerve fiber. The electrical energy is then 10^{-4} erg and of the order of 10^{8} times larger than the energy of a photon in the visible spectrum. The estimate of the energy in a nerve pulse could be in error by several orders of magnitude without altering the major conclusion that an extremely high gain process is needed immediately at the retina before the energy of a few photons can give rise to a nerve pulse.

The progressive reduction in gain as the light intensity is increased is strikingly evident in Hartline's data showing the slow logarithmic increase in frequency of nerve pulses with increasing light intensity. In particular, the ratio of frequencies is only 10 for a ratio of light intensities of 10^4. This means a reduction in gain of 10^3.

While the particular chemistry of the gain process is not known, there is little choice but to expect some form of catalytic action. A photon absorbed by a molecule of sensitive material (rhodopsin) causes a configurational change in the molecule. The ensuing steps whereby the excited rhodopsin catalyzes some ambient biochemical material have yet to be unraveled. However, it is reasonable to expect that the catalytic gain would decrease as the intensity of light or number of excited molecules increased simply because the amount of material to be catalyzed per excited molecule must decrease. Further, it would be expected that the rate of exhaustion of catalyzable material (light adaptation) would be fast compared with its rate of regeneration (dark adaptation). Hence, the common observation that light adaptation occupies only a fraction of a second while dark adaptation may take as long as thirty minutes.

2.9. Visual Noise

If, as we have continued to emphasize, our visual information is limited by random fluctuations in the pattern of incident photons, surely these fluctuations must be visible. Yet this is not the common experience, at least not at normal levels of illumination. We can only conclude that at each light level the gain is delicately adjusted to make the photon noise barely detectable or barely undetectable. If the gain were set higher than this level, the noise would be prominent and annoying without yielding further information. If the gain were set lower than this level, there would be a loss of information. The familiar parallel is (or should be) the adjustment of the gain control knob on a television receiver so that the noise is at the threshold of visibility.

While photon noise is not readily apparent at normal ambient light levels, the writer is quite convinced from his own observations that, at levels around 10^{-5}–10^{-4} foot-lambert, a uniformly lit wall takes on the same fluctuating grainy appearance that one sees in a noisy television picture. Moreover, the visibility of this noise is a strong function of the state of excitement of the observer. A convenient time and place for observation is that just before falling asleep. If, in the course of these observations, a stray sound occurs in the house, portending an unexpected and unwelcome visitor, the flow of adrenalin is momentarily enhanced and, at the same time, a marked increase in the visibility of noise takes place. It is easy to speculate that the mechanisms for self-preservation under these conditions cause the gain of the visual process (indeed the amplitude of all of the sense inputs) to be raised to the level that all of the information is surely being apprehended and, thereby, to the level that the noise is readily apparent.

All of these observations are, of course, necessarily subjective. De Vries is one of the few observers, in addition to the writer, who has hazarded his comparable observations in print. However, many of the writer's colleagues have informally reported similar experiences.

The noise patterns described above are clearly to be associated with the incoming stream of photons since they are absent in those parts of the scene that are "completely black." The presence of some areas of light in the scene is sufficient to define a setting of the gain control such that other much darker areas appear totally black.

If, on the other hand, one is immersed in a completely dark room or if the observer's eyes are completely shielded, the visual impression is not that of a uniformly black field. Rather, one sees a series of faint, shifting, gray patterns already frequently reported in the early literature under the name of "eigenlicht," that is, something generated *within* the visual system. Again, one is tempted to rationalize these observations by saying that since there is no real light pattern to guide the setting of the gain control, the latter is set as high as possible in a search for objective visual information. This high setting is sufficient to reveal the system noise itself which may arise from thermal excitation processes at the retina or may be generated any place in the higher reaches of the nervous system.

A final speculative note concerns the enhanced visual sensations that are reported to accompany various hallucinatory drugs. It would appear highly likely that these drugs produce their effects by increasing the magnitude of gain of the high-gain amplifier located in the retina itself. As we have already noted, a tense or apprehensive emotional state produces a large increase in the amplification.

2.10. Afterimages

The existence of a gain control mechanism at the retina offers an obvious explanation for the variety of observations in which one looks at a bright object and then shifts his gaze to a neutral gray wall. A transient complementary image is seen. For example, a bright black and white scene will yield a transient photographic negative of the original. A bright red object will yield its complementary color, green. In each case, the gain in that part of the retina on which the bright pattern falls has been reduced so that when the retina is exposed to a uniform gray surface, those previously bright areas are transmitted at a lower signal level and appear darker than the surround. The green afterimage of a bright red object shows that the gain mechanism is not only locally variable on different areas of the retina, it can also be set independently for the three color channels at the same area. In this case the red gain was momentarily reduced to reveal the complementary-color image on the neutral gray wall.

It is worth noting that afterimages are not necessarily always negative. If one shields his eyes while facing a bright window, momentarily exposes his eyes to the window, much like a photographic shutter, and is careful to completely shield them again, the afterimage may persist for seconds or even minutes and (at least at the beginning) is clearly a positive afterimage. The positive afterimage is a natural expectation based on the finite decay time for any photoexcitation process in a solid. Since the eye is known to store light signals for 0.1 or 0.2 sec, the mean life of its photoexcitation must also be 0.1 or 0.2 sec and must decay over a period of seconds to progressively lower levels which remain visible since the gain continues to increase after the eyes have been shielded. If, during the observation of the positive afterimage, a small amount of stray light is admitted, the afterimage is converted immediately to a negative afterimage for the reasons cited in the previous paragraph. One can go back and forth between positive and negative afterimages as the stray light is either excluded or admitted.

One type of afterimage may have decorated the English language by way of the phrase "going like a blue streak." If one looks at the end of a lighted cigarette in a dark room and moves the cigarette in a circle, the lighted end will be perceived as a finite streak of light owing to the persistence of vision or positive afterimage. Meantime, the cometlike pattern will be deep red at its head and bluish at its tail. The blue components apparently persist longer than the red components of the cigarette's light. These observations run parallel with the observation that a reddish-colored wall will take on a blue cast as the brightness is reduced below about 10^{-3} foot-lambert. Both sets of observations can be understood if one assumes that the gain control for blue is capable of higher values than that for red and thereby permits the blue sensation to be perceived at lower levels of retinal excitation than red.

2.11. Visibility of High-Energy Radiations

The initiation of a visual sensation is an electronic excitation in a molecule. One would expect, therefore, an energy threshold but, also, one would expect in general that higher-energy radiations would continue to excite the electronic transition and remain visible. It is true that, if the visual excitation is a sharp resonance between two electronic energy levels, the higher-energy photons would not be efficient in exciting the transition. High-energy electrons or ions, on the other hand, can excite a broad range of transitions and should be visible since they leave dense trails of excitations and ionizations in their paths. In an earlier discussion of the visibility of high-energy radiation,[R-5] the writer expressed some surprise that no one had yet reported the direct visual observation of cosmic rays.

There is now evidence for the visibility of a wide range of high-energy radiations. First of all, it has been known for some time that the cutoff in the ultraviolet was a result of absorption by the cornea. Those who have had corneal operations which either removed the cornea or replaced it with a more transparent medium can, indeed, see ultraviolet radiation.

The visibility of x-ray radiation had numerous confirmations in the early days of x-rays. The literature in this area abruptly ceased when the damaging effects of x-rays became known. There was an ambiguity in these early observations as to whether the x-rays excited the retina directly or indirectly via generating fluorescence in the vitreous humor. Some more recent experiments under controlled conditions point to direct excitation of the retina as evidenced by the perception of sharp shadows of opaque objects.[B-4]

The visibility of cosmic rays has now had strong support from the recent reports of astronauts who saw streaks and flashes of light when their cabin was darkened.[W-1] There remains the same ambiguity concerning a choice between direct excitation of the retina or the generation of Cerenkov radiation in the vitreous humor. In view of the dense trail of excitations caused by cosmic rays in any solid, it would be surprising if they did not succeed in directly exciting the retina.

2.12. Vision and Evolution

The ability of living cells to count photons or, at least, to make every photon count was developed early in the history of plant life. The quantum efficiency of photosynthesis for red light is estimated to be about 30%.[A-1] In photosynthesis the energy of the photons is used directly to promote certain chemical reactions. It is not amplified. The plant feeds on light but does not, except for heliotropic effects and the synchronization of biological clocks, use it for information.

The utilization of light for information means that a highly sophisticated amplifier must be constructed immediately at the receptor in order to convert the miniscule energy of photons into the considerably higher energy of nerve pulses. Only in this way can the eye inform the muscles or the brain. Such an amplifier must have made its appearance early in the development of animal life since many primitive animals have their habitat in the dark recesses of the world. Hence, the art of counting photons was mastered long before man made his appearance.

The counting of photons was certainly an essential triumph of the evolutionary process. It was also the most sophisticated step in the development of a visual system. It was essential for survival to ensure that all of the available information could be recorded. Once having this assurance, the adaptation of the visual system to the particular needs of the animal would seem to be an easier and a secondary feat.

The adaptations have taken on a great variety of forms. Most of them appear to have obvious reasons. We cite here a random selection of examples only to confirm the close connection between the optical parameters and the life habits of the animal.

The retinal structure of diurnal birds like the hawk[D-3] is several times finer than that of nocturnal animals like the lemur. Clearly, the high-flying hawk in midday has ample brightness to warrant a finer visual resolution and the correspondingly finer retinal structure. Moreover, the hawk can profit from the enhanced visual detail in his search for a vagrant field mouse. The lemur, on the other hand, in his nocturnal habitat is exposed to such a low level of illumination that the photon-noise-limited quality of his visual images is already coarse-grained and merits no more than a coarse-grained retinal structure. In fact, the paucity of light makes it profitable to have a large aperture, *f*/1.0 lens, even though the lens must be optically of poor quality (Fig. 2.7).

The spectral response of the human eye is closely matched

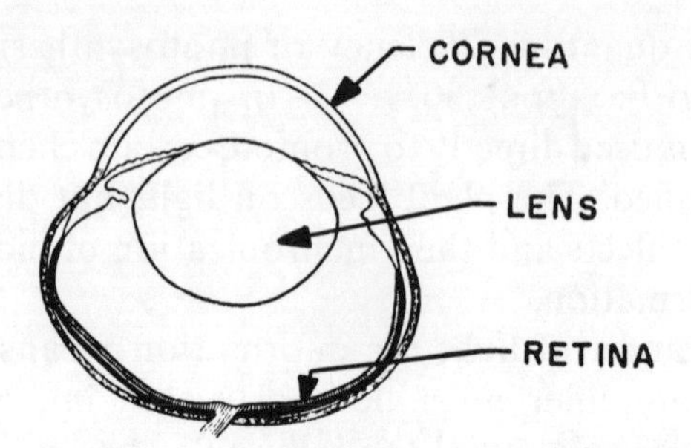

Fig. 2.7. Outline of the lemur eye.

to the peak of the sun's radiation (5500 Å) in daylight. In twilight, the peak of the eye sensitivity shifts toward 5100 Å to match the bluish quality of the light scattered from the sky after the sun has set. One might expect that the eye sensitivity would extend into the red region at least to the point where thermal excitations in the retina would begin to compete with the incoming ambient photons. For example, at the absolute threshold for vision, 10^{-6} foot-lambert, the spectral sensitivity of the eye could be extended to about 1.4 μm before such competition would become significant. It is not clear, therefore, why the eye actually cuts off at 0.7 μm unless the availability of suitable biological materials imposed some constraint.

The storage time of the eye, 0.2 sec, is well matched to the response time, neural, and muscular, of the entire human system. A test of this match is that specially designed television systems with response times of a half second or more become noticeably awkward and frustrating to operate. It is possible that birds have a shorter visual storage time to match their obviously livelier pattern of behavior. Some tangential evidence in this direction comes from certain patterns or series of notes in bird songs which are emitted so rapidly as to sound like chords to the human ear.

There is a close match between the diameter of the rods and cones of the human eye and the diameter of the diffraction disc when the pupil opening is closed down to its smallest diameter of about 2 mm at high light intensities. Various animals have noncircular, slitlike pupils oriented in the vertical direction (e.g., snakes, alligators) or in the horizontal direction (e.g., goats, horses). A vertical slit would yield high definition for vertical lines as limited by lens aberrations and high definition for horizontal lines as limited by diffraction effects. There is ample room for trying to assemble a convincing account of the fitness of these optical parameters to the life habits of the particular animals.

The visual system of the frog is a striking example of adaptation to life habits. The organization of its neural connections is designed to emphasize the movements of delectable (to the frog) flies and to ignore extraneous visual information. Even in the human system we notice a somewhat enhanced sensitivity to flickering lights in our peripheral vision, which clearly can be interpreted as a sentinal system for alerting us to impending danger.

We close this discussion on a somewhat homey note. On the one hand, we have emphasized the close approach of the human eye to the limits imposed by the quantum nature of light. On the other hand, we recognize such phrases in or culture as "to see like a cat" which suggest that the visual sensitivity of the house cat in its nocturnal excursions must significantly exceed our own. It is not altogether unreasonable to reconcile these two statements by noting that if we chose to walk around at night on all fours, we would have the same apparent facility to negotiate obstacles as does the cat.

2.13. Summary

The quantum efficiency of the eye ranges from about 10% at low lights to a few percent at high lights. The total visual range extends from 10^{-7} foot-lambert at the absolute threshold to 10^4 foot-lambert in full sunlight.

A biochemical amplifier with a gain, probably in excess of 10^6, exists at the retina in order to convert the minute energy of the incident photons into the considerably higher energy of the optic nerve pulses. The gain of the amplifier is variable and decreases toward high light levels. The variability of the gain accounts for the phenomenon of dark adaptation and a number of afterimage effects.

The visual systems of human and animal species show strong evidence of evolutionary origin and adaptation.

2.14. References

A-1. W. Arnold, An electron–hole picture of photosynthesis, *J. Phys. Chem.* **69,** 788–791 (1965).

B-1. H. B. Barlow, Measurements of the quantum efficiency of discrimination in human scotopic vision, *J. Physiol.* **160,** 169–188 (1962).

B-2. E. Baumgardt, Mesure pyrométrique du seuil visuel absolu, *Opt. Acta* **7,** 305–316 (1960).

B-3. H.R. Blackwell, Contrast thresholds of the human eye, *J. Opt. Soc. Am.* **36**, 624 (1946).

B-4. H. Bornschein, R. Pape, and J. Zakovsky, Sensitivity of the retina to x-rays, *Naturwiss.* **40,** 251 (1953).

B-5. E.M. Brumberg, S.I. Vavilov, and Z.M. Sverdlov, Visual measurements of quantum fluctuations, *J. Phys.* (*U.S.S.R.*) **7**, 1 (1943). These authors estimated that 25–50 photons are used by the eye in forming its threshold sensations.

C-1. P. W. Cobb and F. K. Moss, The four variables of visual threshold, *J. Franklin Inst.* **205**, 831 (1928).

C-2. J. P. Connor and R. E. Ganoung, An experimental determination of visual thresholds at low values of illumination, *J. Opt. Soc. Am.* **25,** 287 (1935).

D-1. R. L. de Valois, I. Abramov, and G. H. Jacobs, Analysis of response patterns of LGW cells, *J. Opt. Soc. Am.* **56,** 966–977 (1966).

D-2. H. de Vries, The quantum character of light and its bearing upon threshold of vision, the differential sensitivity, and visual acuity of the eye, *Physica* **10,** 553–564 (1943).

D-3. S. R. Detwiler, The eye and its structural adaptations, *Am. Scientist* **44,** 45 (1956).

H-1. H. K. Hartline, Nerve messages in the fibers of the visual pathway, *J. Opt. Soc. Am.* **30,** 239 (1940).

H-2. S. Hecht, The instantaneous visual thresholds after light adaptation, *Proc. Nat. Acad. Sci.* (*U.S.*) **23**, 227 (1937).

H-3. S. Hecht, The quantum relations of vision, *J. Opt. Soc. Am.* **32,** 42 (1942). This author estimated that 7 photons are used by the eye in forming its threshold sensations.

H-4. S. Hecht, S. Shlaer, and M. H. Pirenne, Energy quanta and vision, *J. Gen. Physiol.* **25,** 819–840 (1942).

J-1. R. Clark Jones, Quantum efficiency of human vision, *J. Opt. Soc. Am.* **49**, 645–653 (1959).

R-1. P. Reeves, The response of the average pupil to various intensities of light, *J. Opt. Soc. Am.* **4,** 35 (1920).

R-2. G. T. Reynolds, Princeton Univ., private communication.

R-3. A. Rose, The sensitivity performance of the human eye on an absolute scale, *J. Opt. Soc. Am.* **38,** 196–208 (1948).

R-4. A. Rose, Quantum and noise limitations of the visual process, *J. Opt. Soc. Am.* **43,** 715 (1953).

R-5. A. Rose, "Quantum Effects in Human Vision," in *Advances in Biological and Medical Physics,* Vol. 5, pp. 211–242 (1957), Academic Press, New York.

R-6. J. E. Ruedy, private communication.

R-7. W. H. Rushton, Rhodopsin density in the human rods, *J. Physiol.* **13,** 30–46 (1956).

V-1. H. A. van der Velden, Concerning the number of light quanta necessary for a light sensation in the human eye, *Physica* **11,** 179 (1944). This author estimated that 2 photons are used by the eye in forming its threshold sensations.

W-1. G. L. Wick, Cosmic rays: Detection with the eye, *Science* **175,** 615–616 (1972).

General

R. Clark Jones, "Quantum Efficiency of Detectors for Visible and Infrared Radiation," in *Advances in Electronics and Electron Physics,* Vol. 11, pp. 87–183 (1959), Academic Press, New York.

A. Rose, "Quantum Effects in Human Vision," in *Advances in Biological and Medical Physics,* Vol. 5, pp. 211–242 (1957), Academic Press, New York.

O. H. Schade, Optical and photoelectric analog of the eye, *J. Opt. Soc. Am.* **46**, 721–739 (1956).

CHAPTER 3

TELEVISION CAMERA TUBES

3.1. Introduction

At the outset, we recognize an obvious difference between the human visual process and the equivalent electronic camera tube. In the retina, all of the elements transmit their optical information simultaneously to the brain. In the camera tube, the elements are scanned serially at a rate sufficiently high that they appear to the eye to be simultaneous. The camera tube could, in principle, be designed to have all of its picture elements transmit their information simultaneously to the receiver. The resulting system would be cumbersome compared with the conventional scanning systems. There is, however, no fundamental difference in sensitivity to be expected between a simultaneous and a properly designed sequentially scanned system. The scanning process only imposes its own set of technical problems without fundamentally limiting the performance of the system.

The early television camera tubes around 1930 operated best in the midday sun. Recent camera tubes can function comfortably in full moonlight. This is a span of some 10^6 in scene brightness as well as in camera sensitivity. In fact, the ratio corresponds well with the ratio to be expected from ideal photon-noise-limited operation with and without full electrical storage of the light signals. The early Farnsworth dissector tube, for example, sampled the incoming light only during the time of one picture element. For a picture having 10^5 picture elements, this meant that the light was sampled only 10^{-5} of the full time. Modern camera tubes accumulate the information from all of the light all of the time and hence have a sampling ratio of unity compared with the 10^{-5} of the early dissector-type tubes.

The series of camera tubes is reviewed in this chapter with emphasis on the progression of fundamental advances leading to the modern full-storage, photon-counting camera tubes.

3.2. Scanning Discs and Dissector Tubes

One of the earliest systems (Fig. 3.1) for television consisted of a lens which focused the scene to be transmitted on the surface of a spinning disc. The disc had a pattern of holes so arranged

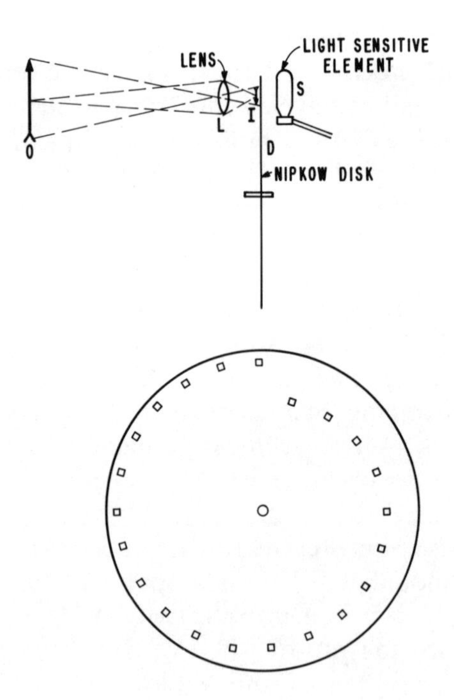

Fig. 3.1. Nipkow disc.

that the scene was scanned in a series of horizontal lines. The light passing through a hole was focused on a photocell whose current was then proportional to the brightness of successive picture elements. The current from the photocell was amplified and transmitted by wire or radio link to a receiver where it modulated the brightness of a lamp. Varous mechanical arrangements were used to cause the image of the lamp to scan a screen in synchronism with the hole in the initial scanning disc. In this way the original scene was reconstructed and the principle of a television system was demonstrated.

The sensitivity of this system was constrained to be at best only a small fraction of the sensitivity of a full-storage system. The dissector, at any one time, transmitted the information passing through one of the small holes and discarded the remainder. The size of the hole defined the size of a picture element. Hence, for a picture having 10^5 picture elements only a fraction, 10^{-5}, of the light was being used at any time.

Farnsworth[F-1] developed a sophisticated version (Fig. 3.2) of the dissector by electrically scanning an electronic image, of the scene to be transmitted, across a small aperture. Behind the aperture, the collected electrons were fed into an electron multiplier. The electrical scanning made it relatively easy to achieve a high-resolution scanning system. The electron multiplier insured

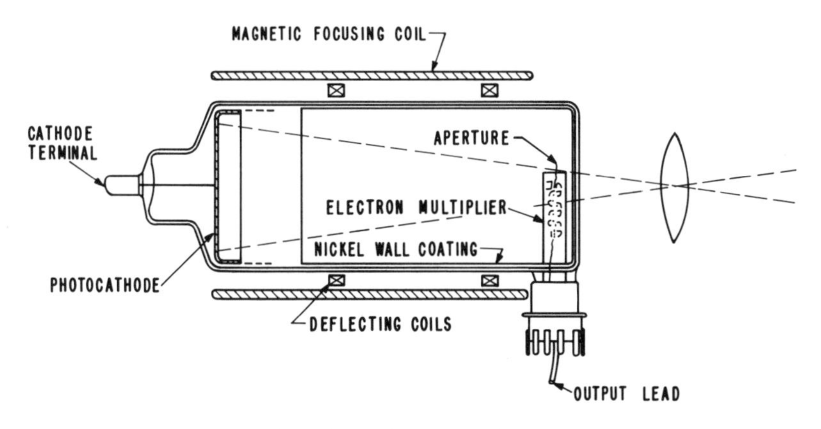

Fig. 3.2. Dissector tube.

that the only limiting noise would be that of the photoelectrons emitted by the photocathode on which the image was focused. If we designate the quantum efficiency of the photocathode by θ, the quantum efficiency of the dissector system is θ/N, where N is the number of picture elements in a single picture.

A dissector tube operating in broad daylight ($\sim 10^3$ foot-lamberts) with an $f/2$ lens and a photocathode quantum efficiency of 10% should provide the eye with a picture whose signal-to-noise ratio at 400 lines and 0.2 sec integration time is approximately 100. This is a good quality picture.

3.3. Iconoscope

A major advance in principle of operation was introduced by Zworykin[(Z-1)] in the form of the iconoscope (Fig. 3.3). Here, the image was focused on a photoemissive insulating surface or target so that, nominally, all of the light was being sensed all of the time. A fine electron beam was used to scan the target. The beam struck the target at about 1000 volts and liberated a low-energy spray of

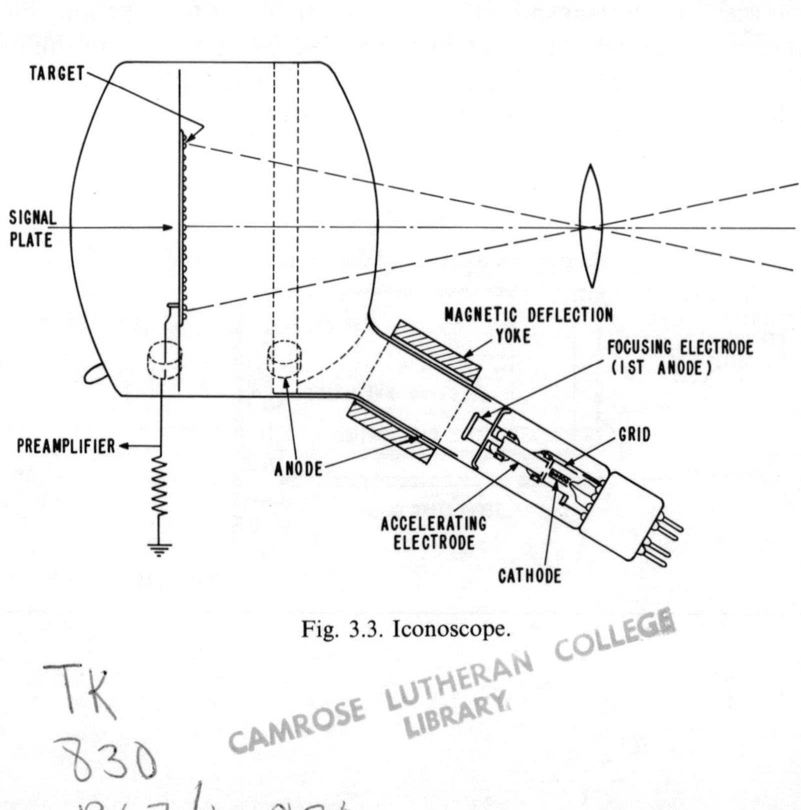

Fig. 3.3. Iconoscope.

secondary electrons. The escape of these low-energy secondary electrons was modulated by the electrical potentials built up at each picture element by the image light. The modulated current to the target was then sensed by a metal electrode (signal plate) and passed on to the first stage of an amplifier as a video signal.

The simple addition of storage to the nonstorage dissector tube should have increased its sensitivity some 10^5-fold and yielded essentially the ideal television camera tube. The iconoscope, however, realized an improvement in sensitivity over the dissector tube of only a factor of about 10. The major reason for not achieving the 10^5-fold increase was that the video signal was amplified by a conventional vacuum triode rather than by an electron multiplier. The vacuum triode introduced its own noise level which was 100–1000 times larger than the noise associated with the signal current in a multiplier tube. Moreover, the amplifier noise was a constant, independent of signal level, whereas the noise currents in the multiplier decreased as the signal decreased.

The iconoscope had other shortcomings which further limited its useful sensitivity. The collection of photoelectrons was inefficient and the redistribution of secondary electrons from the beam over the target led to a further inefficiency of converting the stored video charge into a video signal. This redistribution also brought in a serious shading signal which was present even in the dark, and tended to obscure the low-light portions of a picture.

In one sense, it was remarkable that the inocoscope functioned even as well as it did. One would normally expect the scanning beam to discharge the photoelectrically charged elements of the target by depositing charge of the opposite sign. But in the iconoscope, both the light and the beam charged the target positively. It was only by a complex process of redistribution of the secondary emission from the scanning beam over the remainder of the target that a video signal could be generated at all. This redistribution brought with it an uneven background shading, inefficient discharge, and a spurious interdependence of the signal from different parts of the scene.

Nevertheless, it was the iconoscope that allowed the industrial community to think and plan seriously for the inauguration of a television system.

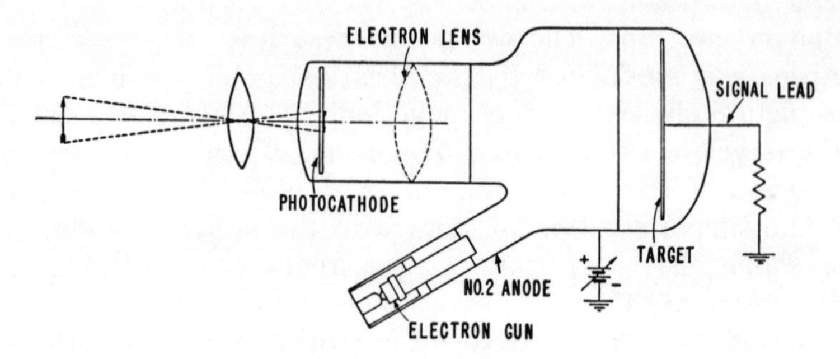

Fig. 3.4. Image iconoscope.

3.4. Image Iconoscope

In the iconoscope, the scene to be transmitted was focused on a a photoemissive, insulating target. In the image iconoscope(I-1) (Fig. 3.4), the scene is focused on a semitransparent conducting photocathode, and the photoelectrons are then focused on the insulating target. The use of the image section increased the sensitivity of the iconoscope by at least a factor of 10. The photocathode could be processed to have a higher sensitivity than the insulator surface. Further, the charge pattern, formed when the electron image struck the target, was enhanced severalfold by secondary emission.

The English counterpart of the image iconoscope, the super emitron, found widespread use in Europe and England for over a decade. In the end, the spurious shading patterns were probably its most serious fault.

3.5. Orthicon

The orthicon(R-3) (Fig. 3.5) introduced the principle of low-velocity beam scanning which has come to be used in almost all subsequent camera tubes. In the dark, the electron beam charges the target surface to the potential of the cathode of the electron gun. The beam approaches the target at close to zero velocity, turns around at the target surface, and returns toward the electron gun. In this way the target is charged uniformly to zero volts (cathode potential), there are no shading signals in the dark, the

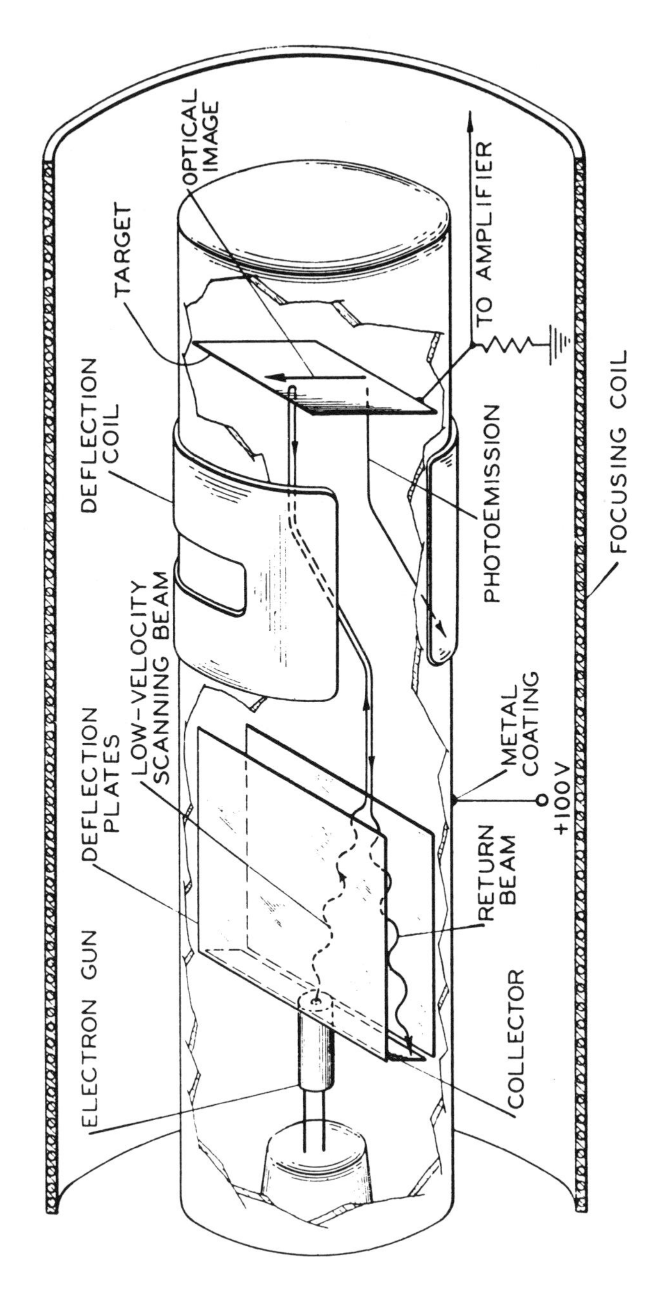

Fig. 3.5. Orthicon.

photoelectrons are efficiently collected, and there is no interaction between different parts of the picture. The beam simply deposits a negative charge at each picture element equal to the positive photocharge accumulated during the previous scanning period. The negative charge is capacitively sensed by the signal plate and fed into the first stage of an amplifier.

The sensitivity of the orthicon was at least a factor of 10 higher than that of the iconoscope. This factor could be significantly larger depending upon how much importance was attached to the spurious shading signals in the iconoscope.

The scanning arrangement shown in Fig. 3.5 is the one used in the earliest orthicons. For the vertical deflection, a longitudinal magnetic field is warped by the transverse field of a pair of deflection coils so that the magnetic line starting from the gun is swept across the target. The electron beam follows the magnetic lines to a first approximation both going to and coming from the target. Second-order effects arise from the local helical motion of the electrons and from the centrifugal force on the electron as it follows a kink in the magnetic field pattern.

The horizontal scanning was carried out by a pair of electrostatic plates, immersed in the magnetic field, deflecting the beam normal to both the electric and magnetic fields. Deflection can also be accomplished by short deflection coils or plates in the absence of an axial magnetic field. An electrostatic lens is then needed near the target to bend the scanning beam so that it always approaches the target normal to the plane of the target. Later orthicons used magnetic deflection for both the vertical and horizontal as in the vidicon.

The CPS emitron was an orthicon with the photosensitive material evaporated on the target through a fine mesh screen. This arrangement led to improved stability of the target at cathode potential even in the face of bright flash bulbs.

3.6. Image Orthicon

The image orthicon[R-4] (Fig. 3.6) exceeded the sensitivity of the orthicon by a factor of 100 or more and brought the sensitivity of camera tubes close to that of photon-noise-limited per-

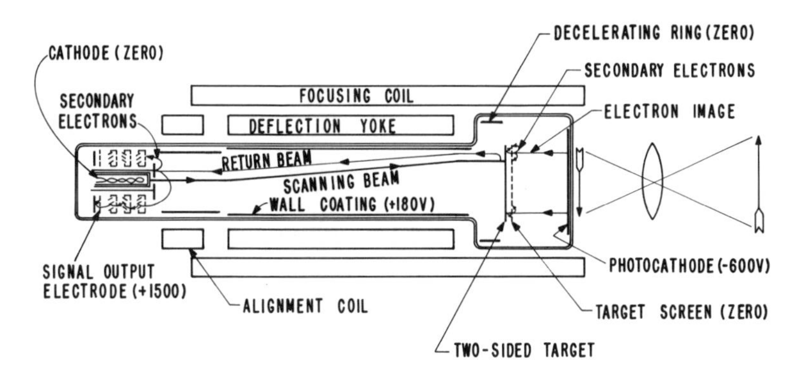

Fig. 3.6. Image orthicon.

formance, at least over a finite range of light intensities. The intrinsic sensitivity (quantum efficiency) of the image orthicon was a close match to that of the human eye over most of the camera tube's operating range.

The scene to be transmitted in the image orthicon was focused on a semitransparent conducting photocathode rather than on a photosensitive insulating target as in the orthicon. The electron image was then focused onto one side of a two-sided target. The two-sided target was a sheet of common window glass (or lime glass) stretched taut on a metal frame by surface tension. The sheet was only a few microns thick. The ionic conductivity of the glass* was 10^{-11} $(\text{ohm}\cdot\text{cm})^{-1}$ and sufficient to allow charges on opposite faces to neutralize each other by conduction in less than a tenth of a second. The glass was thin enough that a charge pattern did not lose its sharp edges by sidewise spreading in less than a second. It was also thin enough that the charge pattern on the image side of the glass appeared just as sharp to the beam as it would have appeared on the scanned surface as in the orthicon.

The scanning beam in the image orthicon was brought back to an electron multiplier surrounding the gun. Under normal operating conditions, the beam current was comparable with the

* This is one of the rare examples of the successful use of ionic conductivity in an electronic device.

photocurrent leaving the photocathode and was modulated by the charges on the target to a depth of about 50%. Hence, the signal-to-noise ratio of the video signal emerging from the electron multiplier was comparable with the signal-to-noise ratio of the photoelectron current leaving the photocathode. In brief, this part of the tube operated close to the photon-noise-limited values of those photons that gave rise to photoemission.

The electron image section represented a tenfold increase in sensitivity over the insulator target surface of the orthicon. The electron multiplier for the scanning beam improved the signal-to-noise ratio by a factor of 10 at normal light levels and by a factor of about 100 at light levels near that of full moonlight. These two features accounted for the gain in sensitivity of 100–1000. Figure 3.7a shows the setup for comparing sensitivities of the image orthicon and 35-mm super XX film; Fig. 3.7b shows their comparative performances down to brightness levels approximating full moonlight. Figure 3.8 shows the first television picture transmitted under actual full moonlight.

A major limitation in sensitivity in the image orthicon arises from the fact that the full-beam current and its attendant shot noise enters the multiplier in the dark areas of the picture. In a

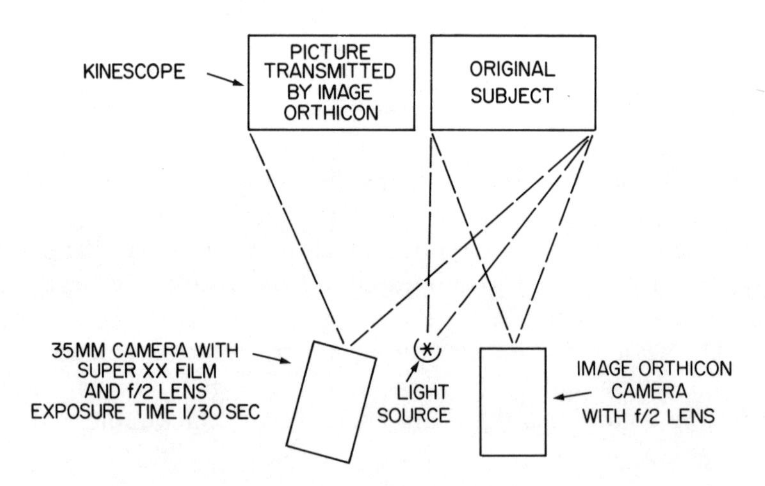

Fig. 3.7a. Setup for comparing sensitivities of the image orthicon and 35-mm super XX film.

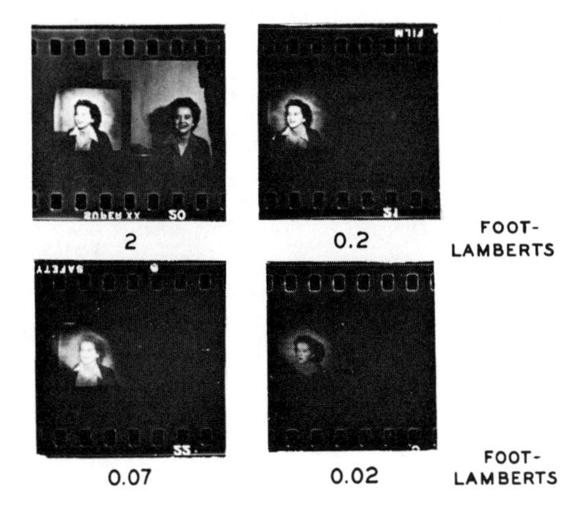

Fig. 3.7b. Comparison of sensitivities of image orthicon and 35-mm super XX film (incandescent light source).

properly designed camera tube, the noise should approach zero as the video signal approaches zero. Two paths were followed to overcome this limitation. One was an attempt to invert the modulation pattern of the beam in the image orthicon. The second comprised a variety of tubes in which the charge pattern of the electron image section was multiplied so that the noise of the stored charge pattern exceeded the noise in the scanning beam and, in more recent examples, also exceeded the noise in conventional television amplifiers.

3.7. Image Isocon

The image isocon[(W-1)] (Fig. 3.9) succeeded, to a significant extent, in inverting the modulation pattern of the image orthicon. In the image orthicon the full beam enters the electron multiplier in the dark. In the image isocon, the beam is largely diverted from entering the multiplier in the dark. In the lighted areas, however, an increasing fraction of the beam enters the multiplier.

The inversion of the modulation pattern was accomplished by an ingenious electron-optical arrangement which distinguished between electrons reflected from the target in the dark areas and

Fig. 3.8. First picture recorded by television in full moonlight. The picture of R.D. Kell was recorded in 1944 on the roof of RCA Laboratories using an early model of an image orthicon camera. The original print from which this copy was made is in the archives of the Niels Bohr Library of the American Institute of Physics.

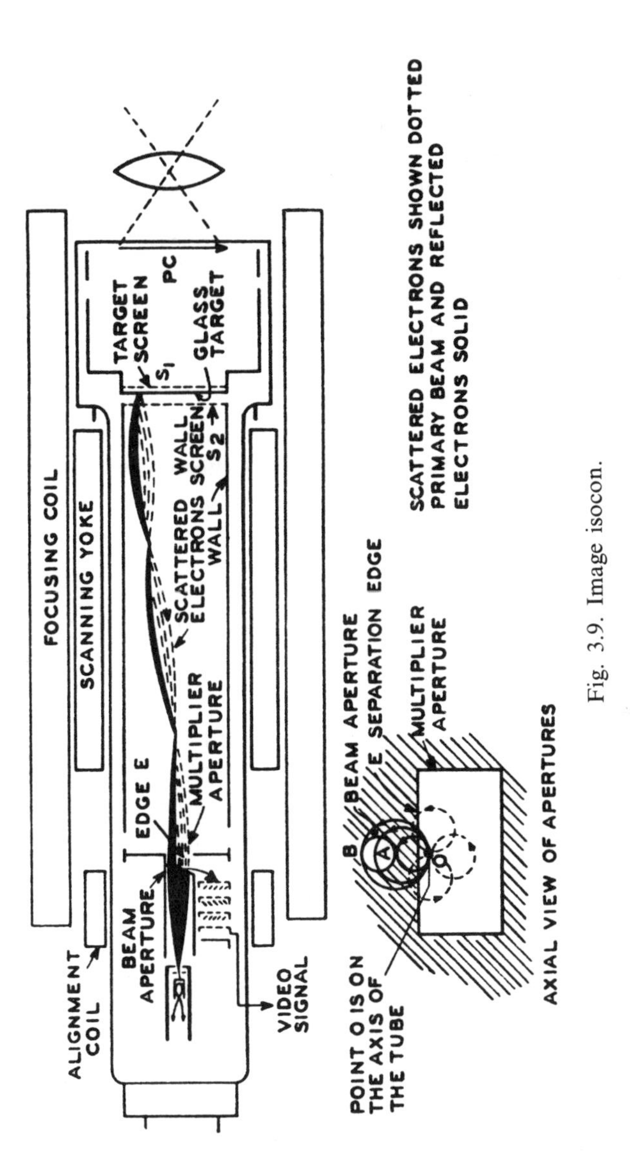

Fig. 3.9. Image isocon.

those scattered from the target in the lighted areas. The reflected electrons remained well focused and returned to a well-defined small spot which was deflected just off the entrance aperture to the multiplier. The electrons scattered back from the lighted areas returned in a diffuse pattern surrounding the finely focused reflected electrons. In this way, about half the scattered electrons entered the multiplier.

The discrimination between reflected and scattered electrons was not complete but it was sufficient to reduce the noise currents in the dark areas by factors of 2 or 3 compared with the noise currents in the high light areas. The isocon made a significant improvement in the sensitivity, noise properties, and low light level performance of the image orthicon.

The isocon principle has been applied to storage-type tubes in which the beam was modulated (deflected) by the potential pattern on the target but did not discharge the pattern. In this way, a charge pattern could be read off for seconds or minutes. The isocon principle has been used, also, in the vidicon type tubes described below.

3.8. Intensifier Image Orthicon

An early realization [M-1] of the second method of overcoming the noise limitations of the scanning beam, by multiplying the electron image until its noise exceeded the noise of the scanning beam, was accomplished by attaching an image intensifier to an image orthicon as shown in Fig. 3.10. An electron from the first photocathode on the left struck the first fluorescent screen at 10 kilovolts and produced several hundred photons which in turn liberated some tens of electrons from the second photocathode located adjacent to the fluorescent screen. In this way, a two-stage intensifier tube multiplied the initial photocathode current by over a hundredfold relative to the image orthicon by itself. The intensifier image orthicon was able to extend the performance of the image orthicon down to light intensities on the photocathode less than 10^{-8} foot-candle. By comparison, the human eye cuts off at light intensities on its retina of about 10^{-7} foot-candle.

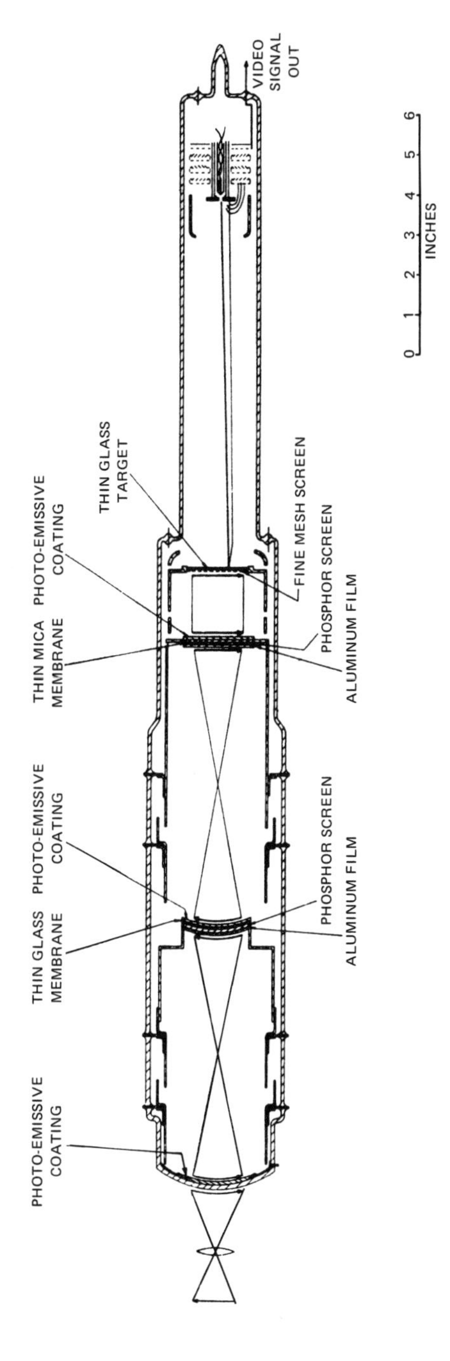

Fig. 3.10. Intensifier image orthicon.

The picture transmitted by the intensifier orthicon was clearly photon-noise limited appropriate to a photocathode quantum efficiency of 10%. The gain in the two-stage intensifier section was, however, still not sufficient to see the trace of single photoelectrons.

3.9. Bombardment-Induced Conductivity

A number of investigators have shown that the conductivity of a thin insulating film could be considerably enhanced by bombarding it with high-velocity electrons whose energy was sufficient to penetrate the film. In fact, the current through the insulating film can easily exceed by a hundred or more times the bombarding current. An early demonstration of this type of multiplication was published by Pensak[P-1, P-2] using a 10 kilovolt writing beam and a 1 kilovolt reading beam. He was able to show hundredfold gains both for targets in which the high-energy beam penetrated the full target thickness and for amorphous selenium where the penetration was only a small fraction of the target thickness. (The latter phenomenon had already been demonstrated by Weimer[W-2] for strongly absorbed light in the selenium vidicon). A similar arrangement with similar results using a lead oxide target was reported[B-1] by Philips at the 1965 London Conference

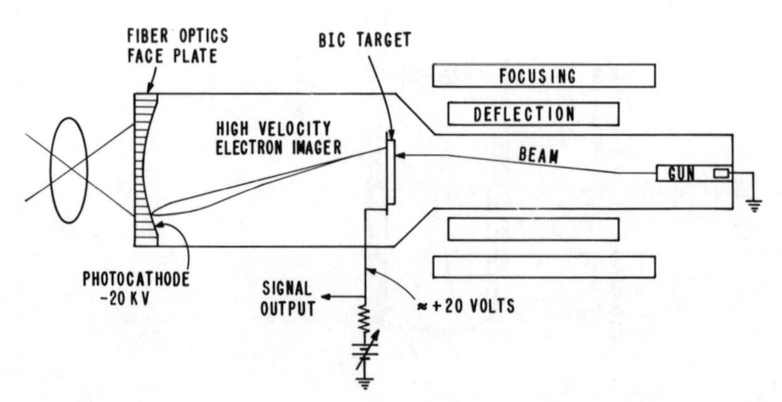

Fig. 3.11. Image orthicon format using bombardment-induced-conductivity target.

on Photoelectronic Imaging. One form of target that reached the level of commercial acceptance was developed by Westinghouse. The tube was known as an SEC (secondary-emission conductivity) tube[(G-1)] and used a target of porous alkali-halide material. The high-energy image electrons produced secondary electrons in the target both in the volume of the alkali halide material and in the pores themselves. The target was operated at relatively low voltages to emphasize the latter contribution. Gains in the order of 100 were achieved.

Finally, the most recent and successful of this class of tubes has been the silicon intensifier tube,[(R-1)] in which the target is a thin slab of silicon about 15 μm thick. The scanned surface has has an array of back-biased *p–n* junctions. The image side of the target is fieldfree *n*-type silicon. The electron–hole pairs, generated by the electrons from the photocathode, diffuse from the fieldfree image surface to the *p–n* junctions at the scanned surface where they are split by the junction field so that the holes reach the scanned surface. With 10 kilovolts on the photocathode, the primary photocurrent generates a thousandfold enhanced current in the target.

A further 30-fold enhancement is achieved by the I–SIT (Intensifier–Silicon Intensifier Target) camera tube, in which the SIT tube is coupled by fiber optics to a one-stage image intensifier tube.

The net effect in all of these bombardment-induced-conductivity arrangements is to extend the operating range of the image orthicon to lower light levels. Note that the intrinsic sensitivity of all these tubes, including the image orthicon and image intensifier orthicon is the same, namely, the quantum efficiency of the photocathode, about 10%. They differ primarily in how far their range of operation extends toward low light levels.

The intrinsic quantum efficiency of 10% matches that of the human eye and, accordingly, would reproduce but not exceed the performance of the eye at low light levels providing the lens opening of the camera tube did not exceed the 8-mm pupil opening of the dark-adapted eye. Since it is normal practice to use lenses with diameters of an inch or two, camera tubes surpass the human eye by the ratio of their respective lens areas, namely, by factors of 10 or more. In principle, this difference in performance should vanish if the human observer used night glasses with their correspondingly larger lens openings.

3.10. Vidicons

All of the camera tubes discussed thus far make use of external photoemission for which the quantum efficiency has not exceeded about 10%. A parallel effort(R-2) was initiated around 1948 to make use of photoconductivity for which a quantum efficiency of 100% could be readily expected. This effort led to the class of tubes known as the vidicon(W-4) (Fig. 3.12). The scanning arrangement is the same as that used for the orthicon, namely, a low-velocity scanning beam focused and deflected in a uniform magnetic field. The target in the vidicon is a photoconductive rather than a photoemissive insulator. In either case, the effect of light is to generate positive charges on the scanned surface which are discharged by the scanning beam.

The initial effort yielded the two fundamental types of targets, those having ohmic contact to the signal plate, as exemplified by the photoconductor Sb_2S_3,(F-2) and those having blocking contact to the signal plate, as exemplified by amorphous selenium.(W-3) Both types have close to 100% quantum efficiency. The blocking

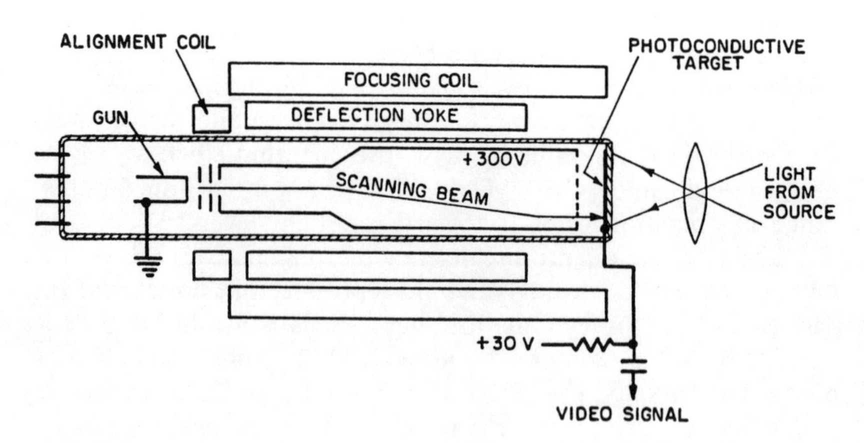

Fig. 3.12. Vidicon.

contact leads to unity gamma (signal proportional to light intensity), to vanishingly small dark currents, and to fast response times. The ohmic contact makes possible gammas that are less than unity (signal proportional to a power of the light intensity less than unity). The ohmic contact, however, requires dark currents of the same magnitude as the video signal if quantum efficiencies near 100% are to be realized. Also, the tubes with ohmic contacts tend to have longer response times than those with blocking contacts. The reasons for these statements will be discussed below in the chapter on photoconductors.

For the present purposes we note only that a number of materials have been found to behave like the amorphous selenium, that is, like materials with good blocking contacts. These include lead oxide,[(H-1)] cadmium sulfide,[(F-2)] gallium phosphide,[(S-3)] cadmium selenide,[(F-2,S-2)] and *p–n* junctions in silicon.[(C-2)] All of these materials have close to 100% quantum efficiency for strongly absorbed light. Lead oxide, as in the plumbicon, is currently the most widely used material for color cameras.

By virtue of the high quantum efficiency of the photoconductor, the vidicons are normally used without an electron multiplier on the return beam. The signal is normally taken out at the target. This leads to greater freedom from spurious signals such as shading introduced by the electron multiplier. The noise of the amplifier connected to the target, however, is some 10 or more times that in the return beam and, hence, reduces the sensitivity correspondingly.

Multipliers have been used on vidicons for special purposes, to achieve either higher sensitivity or, by virtue of the smaller beam currents, higher resolution (see Chapter 5, reference S-1). The multiplier can, of course, be either of the image orthicon type or image isocon type.

Also, as was described in the previous section, an electron-image arrangement can precede the target so that bombardment-induced conductivity can enhance the target current by as much as a thousandfold. In the case of the silicon intensifier tube, photon-noise-limited performance can be obtained down to light levels below starlight, that is, below 10^{-5} foot-lambert. The quantum efficiency is then limited by the photocathode to about 10%.

A perfect low-light camera tube would be achieved if, in the vidicon target, the photoexcited carriers could be made to generate, by impact ionization under high fields, some thousands of secondary electrons. Under these conditions, single photons would be detectable using an electron multiplier on the return beam. The tube would then have 100% quantum efficiency and be able to operate at arbitrarily low light levels. While multiplication by impact ionization is in principle possible and has been observed in some semiconductors, it is extremely difficult to achieve under control for the higher-bandgap materials (i.e., insulators) used in the vidicon. This problem will be discussed further in Chapter 9.

3.11. Solid-State Self-Scanned Arrays

In all of the camera tubes discussed thus far, an electron beam is used to scan a charge pattern on a target. This means, of course, that a vacuum tube is essential. A highly promising alternative approach carried on in the last decade has shown that it is feasible to scan the elements of a photoconductive target by circuit means alone. Weimer[(W-5)] and his co-workers have shown that a 256-line (6×10^4 picture elements) can be addressed or scanned at television rates by a matrix of leads so arranged (Fig. 3.13) that each picture element lies at the intersection of a horizontal and a vertical lead. The 6×10^4 picture elements were accessed by 5×10^2 leads – 256 for the vertical and 256 for the horizontal lines. The total number of leads to the device was reduced from 512 to 64 by use of integrated decoder circuits.

A significant improvement in the method of scanning was suggested in 1969 by Sangter and Teer[(S-1)] using a method called "a bucket brigade" and by Boyle and Smith[(B-2)] in 1970 using a method called "charge-coupling." In both cases, the charge pattern, stored in a line of picture elements, can be stepped across to one edge of the target at video rates and fed into the first stage of a video amplifier built into the same silicon chip on which the picture elements are located (Fig. 3.14). Successive lines of a square array can be scanned off in this fashion.

The self-scanned arrays have the obvious virtues of compact-

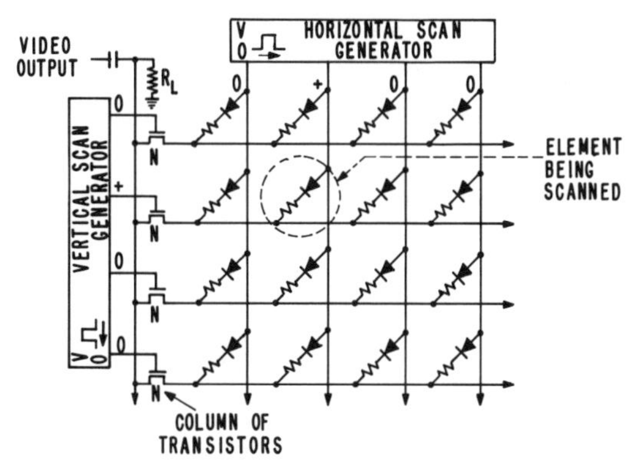

Fig. 3.13. Equivalent circuit for solid-state self-scanned array of sensors (Weimer *et al*).(W-5)

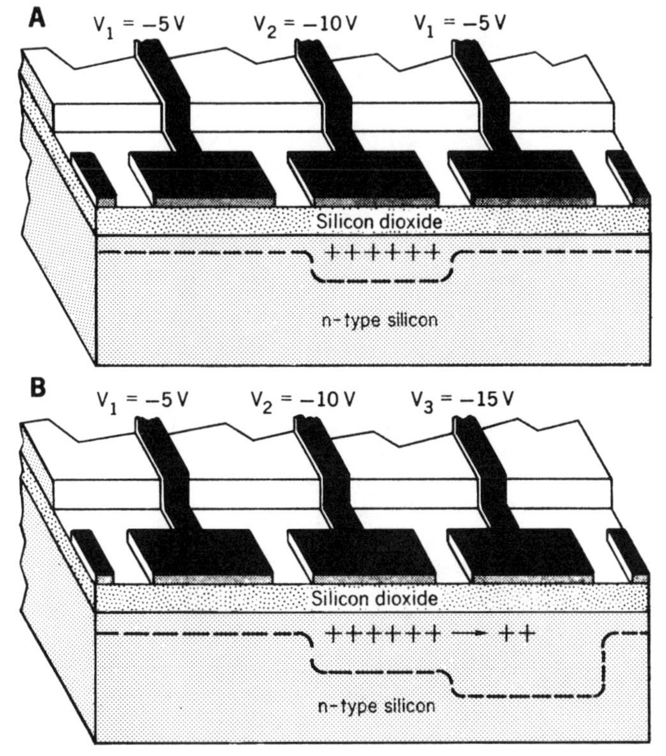

Fig. 3.14. Principle of charge transfer in a charge-coupled sensor (Boyle and Smith). (B-2)

ness, long life, and low power, characteristic of other purely solid-state devices (Fig. 3.15). They have also the fundamental advantage that the first stage of the video amplifier can be built so that its capacitance is 10^{-2}–10^{-4} times that of the usual video amplifier for a vidicon. Since the amplifier noise varies as the inverse square root of the input capacitance, this means a ten- to a hundredfold improvement in sensitivity. To realize the increased sensitivity, it is still necessary to minimize the contributions to noise arising in the scanning process from trapped charge at each picture element.[C-1]

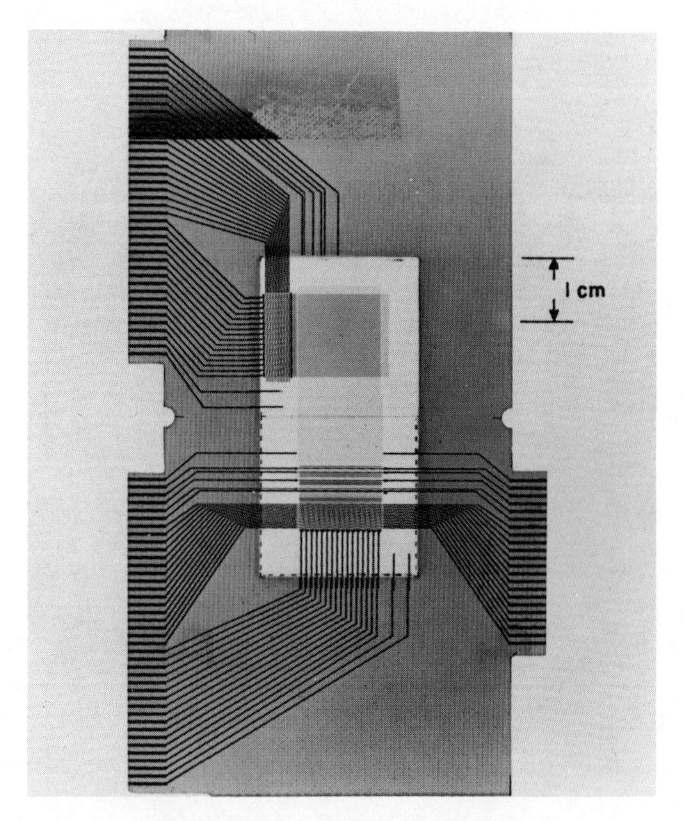

Fig. 3.15. Photograph of complete 256 × 256 element integrated thin-film sensor deposited on two glass substrates mounted on a printed circuit board (Weimer *et al*).[W-5]

A further improvement is, in principle, possible if the first stage is an MOS-type transistor and if it is operated with a floating gate. In the floating-gate mode, the charge of a picture element is placed on the gate and the gate is allowed to float electrically for a time of one picture element. During this time the charge on the gate can induce a much larger charge to flow through the channel of the MOS. After one picture-element time, the charge is removed from the gate and is replaced by the charge of the succeeding picture element. The fact that the gate is floating and not connected to a resistor avoids the usual thermal noise of the resistor. Moreover, if the noise current in the channel of the MOS can be confined to its fundamental thermal value, that is, if the excess noise, called $1/f$ noise, can be sufficiently reduced, it is, in principle, possible to detect single electron charges arising from single photons. The operation of the MOS in the floating-gate mode is essentially equivalent to that of a photoconductor and will be discussed in Chapter 7.

In summary, the charge-coupled image sensor has the formal possibility of achieving the ultimate in performance—a photon counter with 100% quantum efficiency.

3.12. Summary

Television camera tubes have in the span of the last fifty years achieved a millionfold increase in sensitivity.

Modern camera tubes operate with a quantum efficiency between 10 and 100%. Their performance exceeds that of the human eye both in quantum efficiency and in the ability to operate at extremely low light levels in the range of 10^{-7} foot-lambert.

Current and near-future developments in solid-state image sensors will achieve the same high performance in highly compact, low-power, and low-cost formats.

3.13. References

B-1. P.H. Boerse, "Electron Bombardment Induced Conductivity in Lead Monoxide," in *Advances in Electronics and Electron Physics*, Vol. 22A, pp. 305–314 (1966), Academic Press, New York.

B-2. W.S. Boyle and G.E. Smith, Charge-coupled semiconductor devices, *Bell System Tech. J.* **49**, 587–593 (1970).

C-1. J.E. Carnes and W.F. Kosonocky, Noise sources in charge-coupled devices, *RCA Rev*, **33**, 327–343 (1972). [See also M.F. Tompsett, The quantitative effects of interface states on the performance of charge coupled devices, *IEEE Trans. Electron Devices* **ED-20,** 45–55 (1973).]

C-2. M.H. Crowell, T.M. Buck, E.F. Labuda, J.V. Dalton, and E.J. Walsh, A camera tube with a silicon diode array target, *Bell System Tech. J.* **46,** 491–495 (1967).

F-1. P.T. Farnsworth, Television by electron-image scanning, *J. Franklin Inst.* **218,** 411–444 (1934).

F-2. S.V. Forque, R.R. Goodwich, and A.D. Cope, Properties of some photoconductors, principally antimony trisulphide, *RCA Rev.* **13,** 335–349 (1951).

G-1. G.W. Goetz and A.H. Boerio, Secondary electron conduction (SEC) for signal amplification and storage in camera tubes, *Proc. IEEE* **52**, 1007–1012 (1964).

H-1. L. Heijne, Photoconductive properties of lead-oxide layers, *Philips Res. Repts. Suppl.* **4**, 1–161 (1961).

I-1. H.A. Iams, G.A. Morton, and V.K. Zworykin, The image iconoscope, *Proc. IRE* **27,** 541–547 (1939).

M-1. G.A.Morton and J.E. Ruedy, "The Low-Light-Level Performance of the Intensifier Orthicon," in *Advances in Electronics and Electron Physics,* Vol. 12, pp. 183–193 (1960), Academic Press, New York.

P-1. L. Pensak, Conductivity induced by electron bombardment in thin insulating films, *Phys. Rev.* **75,** 472–478 (1949).

P-2. L. Pensak, Electron-bombardment-induced conductivity in selenium, *Phys. Rev.* **78**, 171 (1950).

R-1. R.L. Rogers, III, G.S. Briggs, W.N. Henry, P.W. Kaseman, R.E. Simon, and R.L. Van Asselt, Silicon-intensifier target camera tube, *Int. Conf. on Solid-State Circuits,* February 18-20, 1970, Univ. of Pennsylvania.

R-2. A. Rose, Photoconductivity in insulators, *RCA Rev.* **13,** 303–305 (1951).

R-3. A. Rose and H.A. Iams, The orthicon, a television pickup tube, *RCA Rev.* **4**, 186–199 (1939).

R-4. A. Rose, P.K. Weimer, and H.B. Law, The image orthicon—A sensitive television pick-up tube, *Proc. IRE* **34**, 424–432 (1946).

S-1. F.L. J. Sangster and K. Teer, Bucket-brigade electronics—New possibilities for delay, time-axis conversion, and scanning, *IEEE J. Solid State Circuits* **SC-4,** 131–136 (1969).

S-2. K. Shimizu and Y. Kiuchi, Characteristics of the new vidicon-type camera tube using CdSe as a target material, *Japan. J. Appl. Phys.* **6,** 1089–1095 (1967).

S-3. R.E. Simon, private communication (1966).

W-1. P.K. Weimer, The image isocon—An experimental television pickup tube based on the scattering of low-velocity electrons, *RCA Rev.* **10,** 366–386 (1949).

W-2. P.K. Weimer, Photoconductivity in amorphous selenium, *Phys. Rev.* **79,** 171 (1950).

W-3. P.K. Weimer and A.D. Cope, Photoconductivity in amorphous selenium, *RCA Rev.* **12,** 314–334 (1951).

W-4. P.K. Weimer, S.V. Forque, and R.R. Goodrich, The vidicon—photoconductive camera tube, *Electronics,* May (1950).

W-5. P.K. Weimer, W.S. Pike, G. Sadasiv, F.V. Shallcross, and L. Meray-Horvath, Multi-element self-scanned mosaic sensors, *IEEE Spectrum* **6**, 52–65 (1969).

Z-1. V.K. Zworykin, The iconoscope, *Proc. IRE.* **22**, 16–32 (1934).

General

L.N. Biberman and S. Nudelman, *Photoelectronic Imaging Devices,* Vols. 1 and 2 (1971), Plenum Press, New York.

R. Clark Jones, "Quantum Efficiency of Detectors for Visible and Infrared Radiation," in *Advances in Electronics and Electron Physics,* Vol. 11, pp. 87–183 (1959), Academic Press, New York.

A. Rose, "Television Pickup Tubes and The Problem of Vision," in *Advances in Electronics,* Vol. 1, pp. 131–166 (1948), Academic Press, New York.

O. H. Schade, The resolving-power functions and quantum processes of television camera tubes, *RCA Rev.* **28,** 460–535 (1967).

H.V. Soule, *Electrooptical Photography at Low Illumination Levels* (1968), John Wiley & Sons, Inc., New York.

P.K. Weimer, "Television Camera Tubes: A Research Review," in *Advances in Electronics and Electron Physics,* Vol. 13, pp. 387–437 (1960), Academic Press, New York.

V.K. Zworykin and G.A. Morton, *Television* (1954), John Wiley & Sons, Inc., New York.

CHAPTER 4

PHOTOGRAPHIC FILM

4.1. Introduction

There is a large variety of photochemical processes whereby the action of light serves to darken the material in which it is absorbed. Many of these processes have no photographic gain in the sense that one photon affects at most one molecule of the absorbing material. The old blueprint papers are of this character. The ordinary, but potentially photographic, process of acquiring a suntan is of the same kind. Other and more useful photochemical processes make use of a much smaller optical exposure that produces only a latent (invisible) image which can then be developed to produce a visible image. The most popular and versatile of these processes makes use of micron-sized silver halide crystals imbedded in a gelatin matrix. The silver bromide is sensitized to various parts of the visible spectrum by the adsorption of a wide variety of organic dyes. The gain of the development process is in the order of 10^9, that is, 10^9 silver atoms are generated per "usefully" absorbed photon.

It is remarkable how often one or two chemical compounds, out of an almost limitless range of possibilities, dominate a given technology for many decades. Barium oxide for thermionic emitters, cesium–silver oxide for photoemitters, silicon for transistors, and selenium and zinc oxide for copying machines are well known examples. Silver bromide has similarly dominated the photographic field for over a century. More often than not, a detailed understanding of why a particular compound should be unique is lacking. And, indeed, the compound is probably not unique and is always in danger of being displaced.

The development of silver bromide emulsions is particularly remarkable because it had to be carried out both literally and metaphorically in the dark. Literally, owing to the sensitivity of the emulsion to visible light. Metaphorically, because much of the research was done in the absence of any well-developed models of the band structure of solids. Even now, a detailed understanding of the electronic processes is still lacking or, at least, subject to argument.

The silver bromide photographic process is a highly sophisticated combination of solid-state electronics and chemical reactions. If photographic film did not exist and someone proposed the task of developing a material which could remain "primed" for months, ready to respond to only a few photons, and then remain "latent" again for months, ready to yield a catalytic gain of 10^9 (during chemical development), there would be few, indeed, foolhardy enough to undertake the risk. Fortunately, the problem did not need to be framed in such dramatic dimensions. The slow inching process of science and invention succeeded in building an edifice whose elaborate design could not have been forseen at the outset.

4.2. Sensitivity and Signal-to-Noise Ratio

Consider a photographic negative that has not been exposed to light. It is essentially a sheet of film consisting of micron-sized silver bromide grains imbedded in a layer of gelatin. If the film is now developed in the customary manner, it becomes uniformly transparent. We ignore for the present the usual small amount of fog present in all films. In the final step, a positive print is made from the negative by shining light through it and on to a relatively insensitive print paper. In this case, the print paper would be fully exposed and yield a uniformly black image corresponding to the original picture—which, by assumption, was devoid of light. It is against this black background that we now introduce a finite amount of light to form the image of some test pattern.

Let the test pattern be a small square patch of light which, when imaged on the film, has a length of side d. During the exposure we assume that N_p photons fell on the patch and caused N_g silver bromide grains to be developed into N_g black grains of of silver. When we now shine light through the film to make a print, we end up again with a black picture everywhere except where the N_g opaque grains of silver were present. These yield N_g white spots on the final print. We assume for the moment that all of the silver grains have the same size. We also recognize that the silver-bromide grains and consequently the developed silver grains are randomly distributed so that the rms deviation in the number of grains per unit area is equal to the square root of the average number. The result is that the signal-to-noise ratio in the test patch in the final print is $(N_g/N_g^{1/2}) = N_g^{1/2}$. The signal-to-noise ratio of the incident photons is $(N_p/N_p^{1/2}) = N_p^{1/2}$. Hence, the signal-to- noise ratio of the final picture is what would have been obtained if only N_g photons had been used and each photon had given rise to a white spot. In summary, the quantum yield of the film is

$$\theta = \frac{N_g}{N_p} = \frac{(\text{signal/noise})^2_{\text{ reproduced picture}}}{(\text{signal/noise})^2_{\text{ incident photons}}} \tag{4.1}$$

This is the same criterion, used in Chapter 2, to evaluate the quantum yield of the human eye. R. Clark Jones[J-1] has applied the criterion to a number of commonly used films to arrive at a quantum efficiency ($100 \times \theta$) of about 1%.

While the intrinsic sensitivity of about 1% is common to a wide range of films, the photographic speed is readily varied by varying the size of the silver-bromide grains. Thus, if we take as a nominal criterion that 100 photons must be incident on a grain to make it developable, the density of photons required to expose a given film is 100/(grain area) photons/cm^2. For a medium range of grain size, the area is 10^{-8} cm^2 and the exposure is 10^{10} photons/cm^2. It follows also that the higher-speed (larger-grained) films have a lower signal-to-noise ratio for a given elemental area on the film. The signal-to-noise ratio is approximately the square root of the number of grains in the elemental area.

Our analysis has thus far assumed a single layer of grains, a random spatial distribution of grains, and a uniform size and sensitivity for the grains. In actual practice, photographic films have a thickness of several grains. The assumed random nature of their spatial distribution is valid and has been confirmed experimentally by the observation that the signal-to-noise ratio is proportional to the diameter of test element, that is, to the square root of the number of grains. Finally, the grains have a significant spread around their mean sizes and mean sensitivities. Sensitivity, as used here, refers to the number of incident photons needed to make the grain developable. The spread in grain size and sensitivity increases the latitute of a film. That is, if all of the grains had the same size and sensitivity, they would all tend to become developable within a narrow range of exposures and would yield an objectionably high-gamma film. (Some special-purpose films are deliberately designed for high gamma.)

The net result of the multilayer thickness and the spread in size and sensitivity is to waste photons.[Z-1] The outer layers, for example, shadow the inner layers from the incident photons. The distribution in sizes and sensitivities means that a given exposure which is sufficient to make the larger and more sensitive grains developable is insufficient for the smaller and less sensitive grains. The photons absorbed by the latter are wasted. Thus, the sensitivities of individual grains are significantly larger than the averaged, operational value of 1% derived from Eq. (4.1). Spencer,[S-3] for example, has analyzed the exposure curves to show that the individual grains are in the order of 10 times more sensitive than their averaged performance in a film would indicate. There have been a number of attempts[B-1, S-1, S-2, S-3, Z-1] to separate out all of the factors that contribute to the deterioration of film sensitivity. While there is not agreement in detail, there is a rough consensus that the two major factors are the multiple number of incident photons required to make any one grain developable and the wide spread in size and sensitivity of grains. The number of incident photons required to make a grain developable can, in principle, be further separated into the number that contribute to forming a silver nucleus and the number that are wasted by being trapped (or by recombination) at sites that make no contribution to the silver nucleus. If all of the grains had the same size and sensitivity, this separation could, as shown by Shaw, be identified in practice by observing the sharpness of transition from white to black as a function of exposure. The magnitude of the photographic gamma measures the sharpness of this transition. (The same type of analysis, see Section 2.6, was applied to the eye to try to determine how many of the incident photons were effective in generating nerve pulses.) The very fact that there is no general agreement on the separation is evidence that the spread in grain size and sensitivity tends to obscure the significance of the separation. There is a similar ambiguity, as shown by Shaw, as to how much of the latitude of film can be ascribed to an actual spread in grain size and sensitivity and how much can be ascribed to the statistical spread that one would expect if each grain required only 2 or 3 effective photons to form a silver nucleus. In fact, he argues that a significant improvement in film sensitivity could be achieved by making the size and sensitivity uniform and by depending on this statistical spread in exposure to yield the latitude.

4.3. Resolution, Signal-to-Noise Ratio, and Effective Passband

From purely geometric considerations one might expect that the smallest picture element of a photographic film would be of the size of the individual grains, much as the limit of resolution of halftone engravings is given by the size of the halftone dots, or the limit of visual resolution is given by the diameter of the rods and cones in the retina. If this were indeed so, the resolution of 35-mm film with micron-sized grains would be in the order of 20,000 television lines corresponding to the 1.8-cm vertical picture height. Comparison with the usual 500-line television pictures would then mean that a passband some 1600 times larger than the usual television passband of 5 MHz (megacycles/sec) would be needed to transmit the information on the film. A qualitatively similar conclusion was, indeed, advanced during the early history of the television system.[K-1]

All of the above would be true if the photographic grains formed a regular array and if they could be partially developed to reproduce a gray scale in the manner of the variable-size black dots of a photoengraving. Since, in fact, the grains are randomly distributed and are of an "all or none" character, they act like the random array of photons described in Chapter 1. In brief, if we selected a picture element to be the size of the grains, the signal-to-noise ratio would be of the order of unity and incapable of

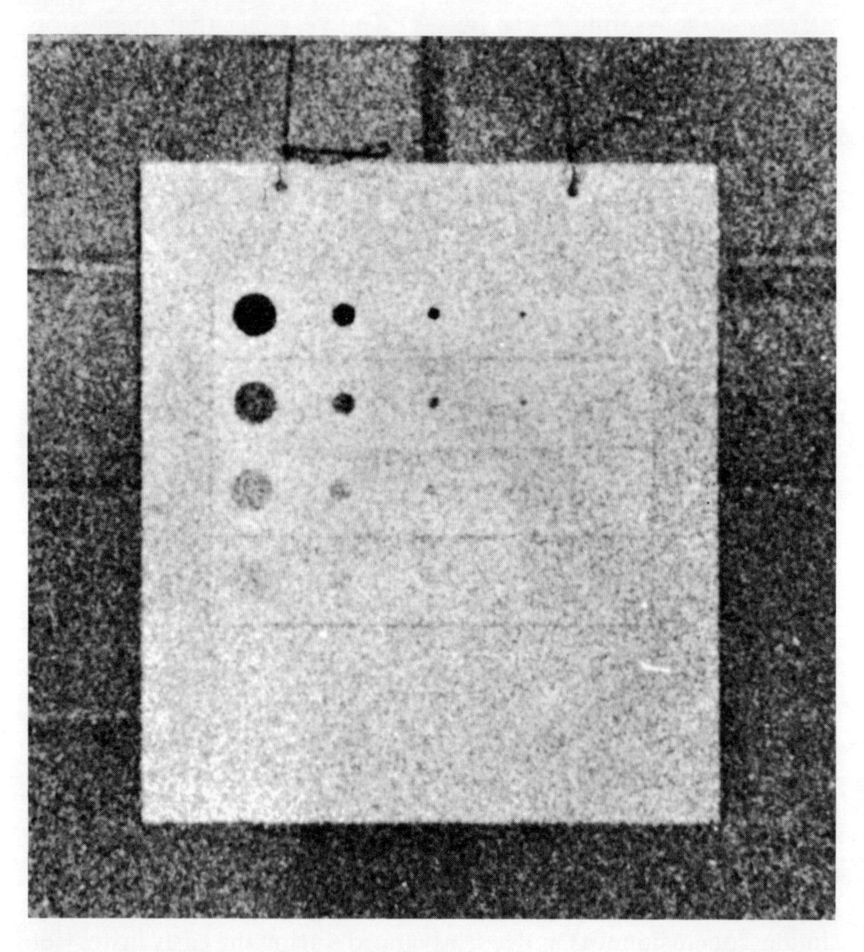

Fig. 4.1. Photograph of the test pattern of Fig. 1.5, on 35-mm super XX film at a low illumination. The purpose of this photograph is to show how the resolution is limited by grain noise. In particular, the low-contrast discs show proportionately poorer resolution in accordance with Eq. (1.3).

transmitting information. The smallest picture element capable of transmitting a single black dot on a white background should have a signal-to-noise ratio of about 5 (see Chapter 1) and should, therefore, contain about 25 grains. Moreover, if we ask for the smallest picture element that can transmit a reasonable contrast such as 10%, the signal-to-noise ratio must be 50 and the picture element must contain 2500 grains. At this point we arrive at a reasonable measure of the resolution of the 35-mm film, namely, about 400 rather than 20,000 television lines.

These considerations determine the normal viewing distance for an audience attending a motion-picture film. At the optimum distance, the visual resolution of the viewer is approximately the 400 lines computed above. At closer distances, the film is noticeably and objectionably noisy. At larger distances, a significant amount of picture detail is visually unresolved. The effect of viewing distance on noise and picture detail is shown graphically in Fig. 5.2b.

The interdependence of resolution, contrast discrimination, and signal-to-noise ratio are brought out in Fig. 4.1. This is a photograph of the test pattern of Fig. 1.5 which consists of a set of discs of varying size and contrast. The test pattern was so arranged that if the relation (Chapter 1)

$$Cd^2 = \text{constant}$$

is satisfied, the demarcation between the visible and nonvisible parts of the test pattern is a 45° line. C is the contrast of the test dot and d is its diameter. Here, again, the strong dependence of resolution on contrast is emphasized. Film resolution deduced from high-contrast test patterns is a poor and grossly misleading measure of the resolving power for low-contrast detail.

4.4. Threshold Properties of Photographic Grains

The quantum efficiency of 1% with which film operates can only be observed near its characteristic exposure of 100 photons per grain area. If the exposure is reduced below this value, the grains do not become developable and the film ceases to function. For this reason, there is a spectrum of films with different grain sizes, extending from the fine-grained, slow-speed copying papers or microfilm to the coarse-grained, high-speed emulsions for use in astronomy.*

The implication here is that at least 2 of the incident 100 photons are needed to form a nucleus for development. If only 1 of the 100 photons were used, the film could still record occasional "hits" down to arbitrarily low light levels. (Actually, the presence of fog, that is, grains that develop with no exposure, would set a finite limit to this excursion.) The fact that 2 or more photons per grain must participate in forming a nucleus is responsible for the rather sharp threshold in exposure, above which the film operates with 1% quantum efficiency and below which it ceases to function.

The actual number of photons or photoexcitations needed to form a nucleus is not known, but is thought to be in the range of 3 or 4 rather than 30 or 40. Each photoexcitation gives rise to a free electron which subsequently becomes trapped at an imperfection of the silver-bromide crystal lying more likely on its surface. The trapped electron attracts a mobile silver ion to its site and forms a silver atom. If several electrons become trapped on neighboring sites so that a cluster of several silver atoms is formed, the cluster can then nucleate the chemical reduction of the entire silver bromide grain to silver. Thus, a chemical gain of some 10^9 is acheived.

The need for more than one photon per grain has, on the one hand, the disadvantage of setting a well-defined threshold exposure below which film ceases to function. On the other hand, it has the important advantage of extending the shelf life of film against thermally induced exposures. It is not sufficient that an electron be thermally excited here and there in separated grains in order to induce a developable exposure. The density of thermally excited electrons must be at least comparable with the density of grains in order that some grains have at least 2 electrons with which to form a developable nucleus.

* The human retina also has a range of "grain" sizes, but these are in physically different areas so that their size increases toward the periphery. The large "grains" are formed by neurally interconnected rods.

A second factor that makes a significant contribution to shelf life is the strong departure of film from reciprocity at long exposure times. The electrons that have been excited, either optically or thermally, and have been trapped at a favorable defect, do not remain trapped indefinitely. They tend to be thermally re-excited into a mobile state where they can eventually recombine into their ground state and annihilate the initial excitation. The data on reciprocity failure show that for exposure times of a few minutes the exposure is twice its optimum value. This can be interpreted to say that at room temperature the trapped electron is released in a time of minutes. Hence, thermal excitations would not be cumulative beyond a period of some minutes, and shelf lives in the order of a year can be achieved. It is not unlikely that cosmic rays eventually terminate the useful shelf life.

4.5. Fog

Except for the presence of fog, the noise in a photographic image approaches zero in the dark parts of the picture. This is the performance to be expected from an ideal picture device. It is achieved by the eye and also by those television camera tubes that have a high degree of image multiplication at, or preceding, the target. It is not achieved by the camera tubes like the vidicon for which the amplifier noise remains constant in the high lights and low lights. Such tubes require more light in order to bring the low lights out of the noise.

The fog in photographic film is the finite density which appears during development, even when the film has been completely unexposed to light. It has somewhat the same effect that a small amount of bias light would have on a perfect or fogfree film. Its magnitude is only a few percent of the high-light density, and its effect is to introduce a small amount of noise in the low-light portions of the photographic image.

4.6. High-Energy Radiations

Photographic film, just as the human eye, is sensitive to a wide gamut of high-energy radiations extending from the few electron volts of ultraviolet light to the millions of electron volts of cosmic rays. Indeed, any radiation that can excite electrons in solids can be recorded by film.

It is particularly interesting to note that the noisiness of x-ray images can exceed the noisiness of the same films exposed to light. The absorption of a 100-kilovolt x-ray quantum can give rise to an electron of the same order of energy. The electron can then penetrate several grains leaving a trail of excitations sufficient to make them all developable. The noisiness of the film is then determined by an effective grain size larger than the original silver bromide grains and equal to the combined size of the several grains exposed by a single x-ray photon.

In another arrangement, due initially to Lallemand,[L-1] the high-energy electrons are generated by light falling on a photocathode. The photoelectrons are then accelerated to some tens of kilovolts before they strike the film. Each electron can make a detectable white speck in the final print. This arrangement has been used by astronomers to record faint stars and nebulae and to extend the photographic process to the counting of individual photons.

In the case of the extremely-high-energy cosmic rays or nuclear particles, photographic emulsions have been used for some decades to record their centimeter-long paths. The length of path and the density of excitation along the path both serve to identify the charge, mass, and energy of the impinging particle. It is only recently, following the observations of streaks of light by the astronauts in their darkened cabins, that the visibility of these same cosmic tracks by the human retina has come to be generally recognized.

4.7. Relative Sensitivities of Film, Television Camera Tubes, and the Human Eye

Figure 4.2 is a rough outline of the comparative sensitivities of photographic film, some representative television camera tubes, and the human eye. Also shown are the approximate ranges of scene brightness over which these various visual systems operate. To make the comparison meaningful, a common exposure

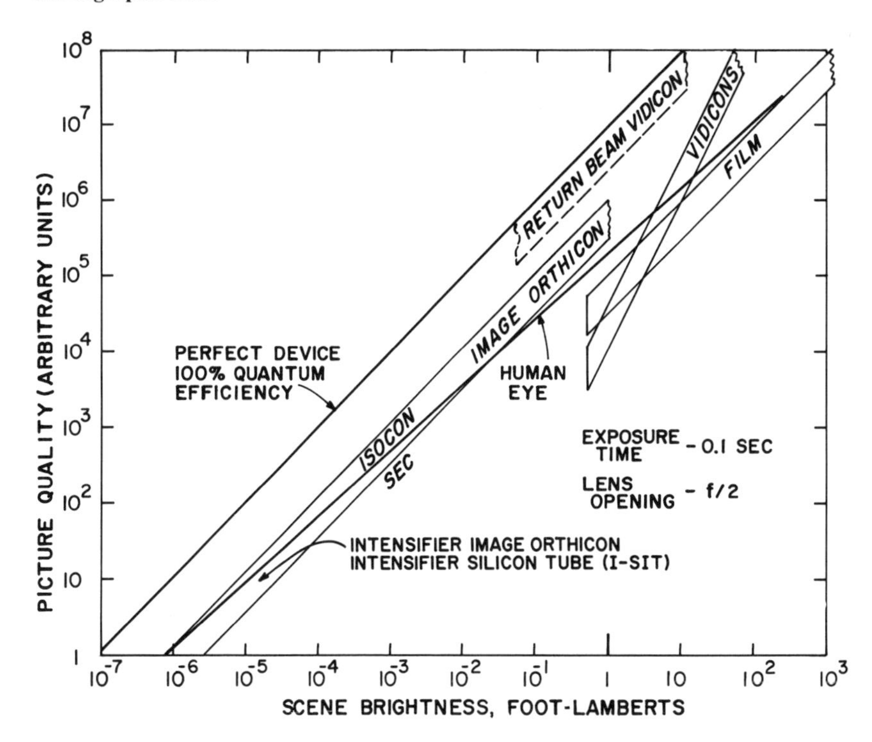

Fig. 4.2. Relative sensitivities of film, television camera tubes, and the human eye.

time of 0.1 sec was assumed together with an $f/2$ lens. The plot has a quantitative meaning on an absolute scale since the curve for an ideal device with 100% quantum efficiency is included. The vertical separation of any of the devices from the ideal curve is a measure of its quantum efficiency.

All of the camera tubes with the exception of the vidicon use photocathodes whose quantum efficiency is about 10% and, therefore, comparable with or greater than that of the eye. These tubes are arranged in order of their ability to detect low-light scenes. Photographic film is plotted along a line of quantum efficiency equal to 1%. The range of scene brightnesses reflects the range in average grain sizes. The curve for the vidicon is steeper than the other curves because it has a fixed amplifier noise. In the limit of high brightnesses the vidicon intersects the ideal curve when the noise in its stored-charge image matches the amplifier noise. At this point it would operate with close to 100% quantum efficiency and a noise level given by that of the photon flux. The intersection is at an unrealistic level of video signal, about 10 microamperes and 50 times higher than its normal video signal. The dotted curve shows the performance to be expected from a vidicon when the signal is taken out through the return beam after going through a multiplier as in the image orthicon [see also reference (S–1) in Chapter 5].

4.8. Summary

The operating sensitivity of photographic film is in the neighborhood of 1% quantum efficiency.

The intrinsic sensitivity of the individual photographic grains is in the neighborhood of 10% quantum efficiency.

The factor of 10 reduction from intrinsic to operating sensitivity is a consequence of the conversion of a digital system to an analog system. The "all or nothing" character of individual grains leads to a distribution of grain sizes and sensitivities in order to achieve a useable latitude in light levels. This distribution, in turn, leads to an underutilization of the photon flux.

Photographic film achieves a range of photographic speeds by use of a range of average grain sizes. The operating sensitivity remains substantially constant. High photographic speed is achieved at the expense of signal-to-noise ratio.

The sensitivity of photographic film is of the order of that of the human eye in the high light range (room light and above) where film is operative. The sensitivity of television camera tubes is of the order of that of the eye at low light levels (twilight and below) and can exceed the eye at high light levels by a factor of 10–100.

4.9. References

B-1. G.R. Bird, R. Clark Jones, and A.E. Ames, The efficiency of radiation detection by photographic films: State-of-the-art and methods of improvement, *Appl. Opt.* **8**, 2389–2405 (1969).

J-1. R. Clark Jones, "Actual and Theoretical Efficiencies of Photographic Film," paper presented at 23rd Annual Conference on Photographic Science and Engineering, May 18, 1970, New York.

K-1. I.J. Kaar, The road ahead for television, *J. Soc. Motion Picture Television Engrs.* **32**, 18–40 (1939).

L-1. A. Lallemand, *Compt. Rend.* **203**, 243 (1936).

S-1. O.H. Schade, An evaluation of photographic image quality and resolving power. *J. Soc. Motion Picture Television Engrs.* **73**, 81–119 (1964).

S-2. R. Shaw, Image evaluation as an aid to photographic emulsion design, *Phot. Sci. Eng.* **16**, 395–405 (1972).

S-3. H.E. Spencer, Calculated sensitivity contributions to detective quantum efficiency in photographic emulsions, *Phot. Sci. Eng.* **15**, 468 (1971).

W-1. G.E. Wick, Cosmic rays: Detection with the eye, *Science* **175**, 615–616 (1972).

Z-1. H.J. Zweig, Theoretical considerations on the quantum efficiency of photographic detectors, *J. Opt. Soc. Am.* **51**, 310–319 (1960).

General

G.R. Bird, R. Clark Jones, and A.E. Ames, The efficiency of radiation detection by photographic films: State-of-the-art and methods of improvement. *Appl. Opt.* **8**, 2389–2405 (1969).

R.W. Engstrom and G.A. Robinson, Choose the tube for low-light-level television, *Electro-Optical Systems Design,* June, 1970.

T.H. James and G.C. Higgins, *Fundamentals of Photographic Theory* (1960), Morgan and Morgan, New York.

R. Clark Jones, "Quantum Efficiency of Detectors for Visible and Infrared Radiation," in *Advances in Electronics and Electron Physics*, Vol. 11, pp. 87–183 (1959), Academic Press, New York.

A. Rose, A unified approach to the performance of photographic film, television pickup tubes and the human eye, *J. Soc. Motion Picture Engrs.* **47**, 273–294 (1946).

O.H. Schade, An evaluation of photographic image quality and resolving power, *J. Soc. Motion Picture Engrs.* **73**, 81–119 (1964).

O.H. Schade, Resolving power functions and integrals of high-definition television and photographic cameras—A new concept in image evaluation, *RCA Rev.* **32**, 567–606 (1971).

H.J. Zweig, Theoretical considerations on the quantum efficiency of photographic detectors, *J. Opt. Soc. Am.* **51**, 310–319 (1960).

CHAPTER 5

COMPARATIVE NOISE PROPERTIES OF VISION, TELEVISION, AND PHOTOGRAPHIC FILM

5.1. Statement of the Problem

A television system is, for the most part, a surrogate for the human eye. The television camera in the studio or at some public event should be seeing what we, as viewers, would see if we were present. Consciously or unconsciously, we judge the quality of the transmitted picture in terms of what our own visual system would have apprehended when standing alongside the camera. For our purposes, this statement has particular meaning for those situations in which the camera is picking up scenes at low light levels, where the information transmitted is limited by the amount of light available. The television picture is then noise-limited as, indeed, our own visual system would be when viewing the original scene. The difference is that the television picture is presented at a high light level so that its noise is readily visible. Our own visual system in viewing the original scene would automatically adjust the gain of the visual process to make the noise only barely visible.

In general, our judgement that a television picture (or motion picture) is, or is not, noisy depends on the relative signal-to-noise ratios of the television picture itself and of our visual image of the television picture. Conventional measures of television noise, however, do not allow an easy comparison of these two signal-to-noise ratios. The conventional measures are based on the scanning process by which the television picture is presented. Hence, the signal-to-noise ratio is computed for the passband of the television amplifier and may or may not be relevant for comparison with the visual image.

The purpose of this chapter is to compare the signal-to-noise ratios of a television system and of photographic film with the signal-to-noise ratio of the visual image in meaningful terms, and to avoid some of the ambiguities frequently present in the literature.

5.2. A Proper Measure of the Signal-to-Noise Ratio in a Television Picture

A proper measure of the signal-to-noise ratio of a television picture, for comparison with the visual image of the picture, should be computed for a given elemental area on the television receiver and for the image of the *same* elemental area on the retina. Furthermore, the signal-to-noise ratio in each case should be computed for the exposure time of 0.1–0.2 sec of the visual process. Normal practice, for example, is to specify the signal-to-noise ratio of a television system for the passband, about 5 MHz, of the amplifier. This value is valid for the smallest picture element, about 1/500 of the picture dimensions, and for one picture time, i.e., 1/30 sec. The conventional value of signal-to-noise ratio can be adapted to areas larger than the smallest picture element by multiplying by the ratio of diameters and to integration times of 0.2 sec. by multiplying by the factor $[(2/10)/(1/30)]^{1/2} = 2.5$. It is this last factor that accounts for the observation that single frames of a television picture or of a motion picture look considerably noisier than the film in motion. The eye normally integrates some six frames of a television picture to yield a corresponding improvement in signal-to-noise ratio.

A not uncommon misuse of the signal-to-noise ratio is to point to the visibility in Fig. 1.6c of one of the larger black discs under conditions when the signal-to-noise ratio for the television system is far less than unity. The latter value applies to the finest picture elements which are about 1/500 of the picture dimensions and which are completely below the threshold of visibility. Those areas of the test pattern which are visible have a signal-to-noise ratio of 5 or greater. Morgan,[M-1] for example, cites the visibility of certain bar patterns when the signal-to-noise ratio of the television system is only 0.1. In the example cited by Morgan, the signal-to-noise ratio of 0.1 characterized the usual picture element whose diameter was about 1/500 of the picture dimension. The observer, on the other hand, was viewing a set of bars and was visually able to integrate not only along the length of a bar, but also over a number of bars. The signal-to-noise ratio of what the observer was detecting was accordingly larger than that of a single picture element by a factor well in excess of 10. Hence, the observer was detecting a pattern whose signal-to-noise ratio was well above unity rather than well below unity.

The same type of error is made when the resolution of a television system (or photographic film) is measured by observing the visibility of small sets of resolution bars. *The fact that a set of bars, whose individual bar width is, for example,* 0.1 *mm, is visible does not mean that a single element* 0.1 *mm on a side will be visible.* The single element is visually smaller than the set of bars and has a smaller signal-to-noise ratio. It will be completely undetectable when the corresponding set of bars is just detectable.

The use of bar patterns to measure the visibility of fine detail is misleading not only for the reasons just outlined, but also for the reason that the bar patterns usually have 100% contrast. As shown in Chapter 1, an element with 50% contrast must be twice as large (in linear dimensions) as an element with 100% contrast in order to be just detectable. An element with 10% contrast must be 10 times as large. For these reasons the conventional use of bar patterns to measure the resolution of a noise-limited system grossly overestimates the visibility of fine, low-contrast detail. The proper test pattern that should be used is that shown in Fig. 1.15. This pattern consists of *single* test elements of varying sizes and varying *contrast*. And it is this type of test pattern that leads to the conclusion that signal-to-noise ratios for threshold detectability must be in the range of 3 to 5.

An example of the magnitude of error provoked by the use of high-contrast bar patterns to measure resolution occurred, as already mentioned, in the early days of television. It was estimated, on the basis of bar-pattern data, that television channels 80 MHz wide would be needed to transmit motion-picture films. These estimates are to be compared with the observations that the quality of film transmissions via the present 5 MHz television channels is frequently observed to be inferior to that of live studio pictures. That is, the quality of the transmitted picture is limited by the quality of the film, and not by that of the television system.

5.3. A Comparison of Arrangements for Noise Reduction

Given a noisy television picture, what are the ways of filtering out the noise, and how do they compare in efficiency?

The simplest way of reducing noise in a television amplifier is to reduce its bandwidth. This, of course, also reduces the ability of the system to resolve fine picture detail. One can also choose to reduce the visibility of noise at the expense of picture detail by backing away from the television screen so that fine detail is not visually resolved. A third method, not generally recognized, is to interpose a neutral filter between the viewer and the screen so that the resolution and contrast discrimination of the eye are reduced. We will show that this is, by far, the most effective noise filter.

There is a fourth method for noise reduction which entails altering the content of the picture itself, and which, accordingly, is in a different category from the three methods just outlined. This method consists of interspersing areas of high contrast throughout the picture. We will discuss the meaning of this procedure separately below.

Of the three methods for noise reduction, reduction in bandwidth is the most ineffective, increased viewer distance loses some picture information; and the use of a neutral filter is the most effective in that no picture information is lost and a realistic or three-dimensional picture quality is added.

To clarify these statements, it is necessary to compare the signal-to-noise ratios of various-sized elements in the television picture with the signal-to-noise ratios of these elements in the visual image on the retina. Fig. 5.1a shows the comparative signal-to-noise ratios of a television picture and of the visual image of that picture under such conditions that the viewer judges the picture to be noisy. The signal-to-noise ratio at the retina and at the given level of picture brightness is larger than the signal-to-noise ratio of the electrical signal.

The signal-to-noise ratio for the television picture increases linearly with the linear size of element considered. We now look at the retinal image of this picture and, for example, count the number of photons falling on the retina for an element of a given size in the picture and for the exposure time of the eye. We then take a fraction, say in the order of 5 %, of these photons corresponding to the quantum efficiency of the eye, and use this number to compute the signal-to-noise ratio of the retinal image. The signal is the number itself; the noise is the square root of the number; and the signal-to-noise ratio is also the square root of the number—that is, the square root of the number of utilized photons. The curve for the signal-to-noise ratio of the retinal image, or eye, was constructed in this way in Fig. 5.1. The eye curve cuts off at small element sizes corresponding to the limit of resolution set by the finite size of rods and cones. It cuts off also at large signal-to-noise ratios, in the order of a few hundred, corresponding to the so called Weber–Fechner limit of about 2 % image contrast. That is,

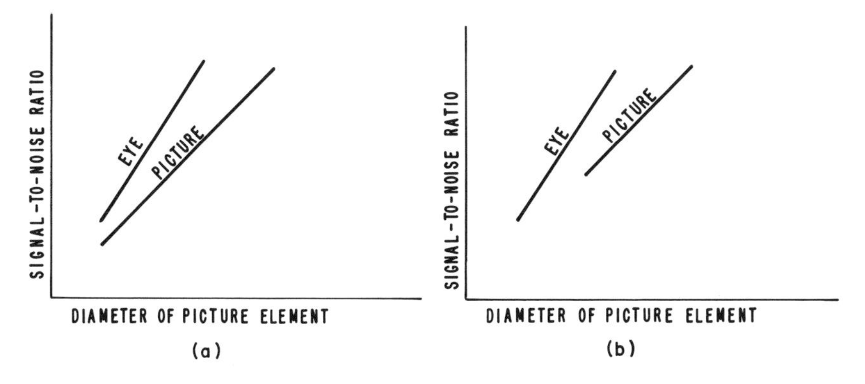

Fig. 5.1. Noise reduction by reduced bandwidth (linear scale).

the visibility of lower contrasts at larger areas is not governed by noise considerations, but by whatever absolute constraints give rise to the Weber–Fechner limit.

Since the eye curve lies above the curve for the television picture, the eye is easily able to see the noise in the television picture. Its discrimination is in this sense finer than the noise texture of the picture.

The total noise in the television picture is related to the total area under its curve. The particular relation is not important—it corresponds to the usual process of integrating noise power over a finite bandwidth. What is significant, is that we can make large reductions in the total noise of the television picture by reducing its bandwidth. Such a reduction is illustrated in Fig. 5.1b by cutting off the finest picture detail at larger element sizes (i.e., fine detail is not transmitted). It is clear from Fig. 5.1b that even though the total noise power in the television channel has been markedly reduced, the noisiness of the picture remains. That is, the eye can still see noise in the coarser (large area) parts of the picture that survived the operation of bandwidth reduction. The appearance of noise has changed from fine grained to coarse, but its visibility remains.

From Fig. 5.1 we conclude that reduction of bandwidth has no virtue. It reduces picture information without altering the noisiness of the remaining picture information. It is necessary, however, to make one qualifying remark. If the passband of the amplifier had initially been much wider than needed to transmit the picture detail, the additional high-frequency noise would tend to reduce the picture contrast in the darker areas of the picture. Black would no longer be black, but a noisy shade of gray. Under these conditions, a reduction of bandwidth does have the virtue of improving the quality of the picture. Amplifiers that selectively reduce the bandwidth in the low lights are able to realize this improvement without loss of picture quality in the high lights. These remarks apply to noise that is added to the picture by, for example, the first stage of a television amplifier. It does not apply to noise that is inherent in the light signal or, in the case of photographic film, to noise that approaches zero as the light signal approaches zero. Photographic film, for example, is recorded at an equivalent passband far in excess of the passband at which it is viewed. If the viewer can resolve elements only as small as 10 microns on a side, and if the lens recording the image on the film were capable of resolving a micron, then the recording passband would be 100 times larger than the viewing passband. This disparity presents no problem because the noise in the film approaches zero in the black parts of the film and a well-defined black level is preserved. A small departure from ideal behavior is produced by the presence of the so-called photographic fog whose density is of the order of 0.02. The effect of the fog is what would be expected from an ideal film on which a small bias light of a few percent of the picture light is always present. The noise accompanying the bias-light exposure prohibits the achievement of a perfect black level.

Fig. 5.2 shows the effect of increased viewer distance on the visibility of noise. The increased viewer distance (Fig. 5.2b) simply introduces a scale factor that shifts the curve to the right by a constant factor. That is, a given area on the retina now corresponds to a larger area of the picture. The effect of increased distance is twofold. Some of the picture information is lost since the eye curve does not extend to the finest detail of the picture. The remainder of the picture, which the eye does see, is now noisefree since the eye curve lies below the picture curve. The discrimination

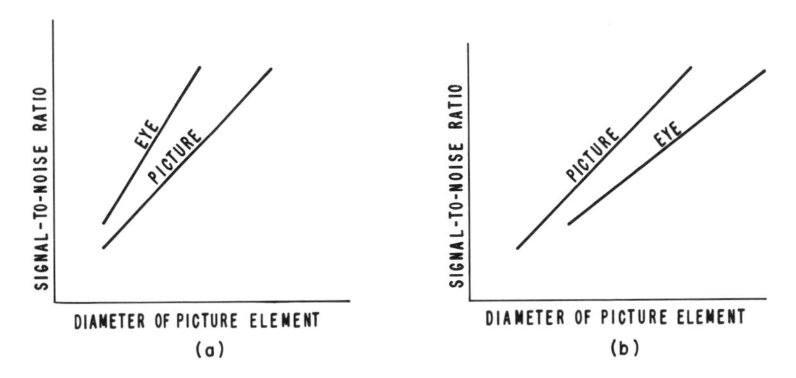

Fig. 5.2. Noise reduction by increased viewing distance (linear scale).

of the eye for the small noise-brightness fluctuations is not good enough to detect them.

Figure 5.3 shows the effect of interposing a neutral optical filter between the observer and the television screen. This time the eye curve is shifted vertically downward (Fig. 5.3b) by a constant scale factor from its initial position shown in Fig. 5.3a to a position lying beneath the curve for the television picture. The effect of the neutral filter is simply to reduce the signal-to-noise ratio at the retina by reducing the number of photons arriving there. Note that this time all of the picture detail is preserved. Only the noise is filtered out, and it is filtered out uniformly for large and small areas of the picture. In place of a neutral filter, the observer can merely squint his eyes or view the picture through a small aperture formed by his fingers in order to reduce the light falling on his retina. In any case, the resultant conversion of a noisy to a noise-free picture is dramatic. One has even the impression that the information content is increased. If so, the real increase of information must be a second-order effect. The impression is largely psychological and comes from the fact that before the neutral filter was interposed the television picture was obviously noisy and looked to be of lower quality than the visual image of real objects in the room. After the neutral filter, the television picture is apparently noisefree and looks to be of the same quality as the real objects. *The eye (or brain) is not aware that the matching pro-*

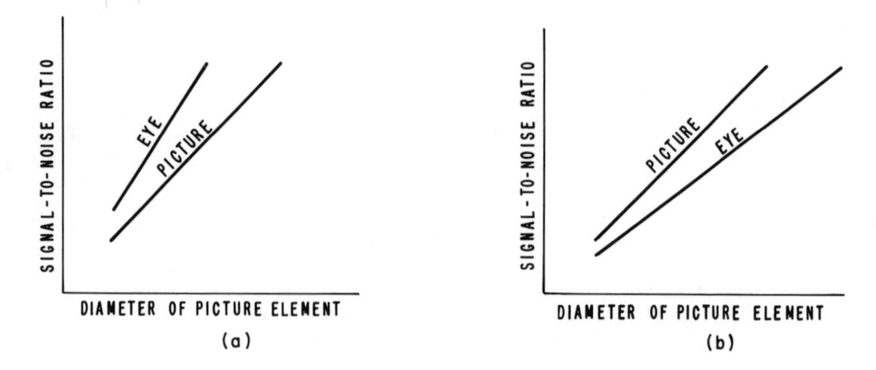

Fig. 5.3. Noise reduction by reduced brightness (linear scale).

cess has been effected by deteriorating the visual image quality of real objects (by reducing the discrimination of the eye) rather than by improving the quality of the television picture.

Another favorable effect of the use of a neutral filter is to increase the apparent contrast of the picture by viewing it at lower light intensity. The logarithmic response of the eye to light intensity leads to a higher photographic gamma at low light levels. It is a common experience to find a white house standing out in high contrast to the landscape in twilight as opposed to its appearance at midday.

Finally, there is a very subtle but very real contribution to the three-dimensional quality of an image that comes from removing all trace of noise. In general, one's sense of three dimensions is largely based on experience rather than on the stereo effect of using two eyes. If this were not so, closing one eye would dramatically convert the image of a real room into a two-dimensional canvas of the room. Retinal images, whether of real three-dimensional objects or of photographs, are necessarily only two-dimensional. The eye tends, therefore, to regard every image as three-dimensional unless there are telltale clues that force it to conclude that it is a two-dimensional canvas or photograph, etc. One of these clues, for example, is the frame of a canvas or the border of a magazine illustration. Viewing the same picture through a tube which excludes the border often makes a striking contribution to the three-dimensional quality of the picture. Similarly, if the fabric of the canvas or the texture of the magazine paper is prominent, the eye has a ready clue for concluding that the picture is two-dimensional. In this sense, noise in a television picture is like the texture of magazine paper. Its presence is the telltale clue that the picture is two-dimensional. Its elimination opens the way for the more natural three-dimensional interpretation.

5.4. Effect of High Contrast on the Visibility of Noise

In the early days of motion pictures, the quality of film tended to be noticeably grainy or noisy. It was recognized empirically that the grainy appearance could be considerably attenuated by avoiding large uniformly lighted areas. A picture that was well broken up by an assortment of bright and dark areas, almost checkerboard fashion, was far less noisy than, for example, a picture with a large gray wall or with a large area of sky.

These observations stem from the fact that the eye has strong overshoots at black–white borders. At the border, the blacks are blacker than black and the whites are whiter than white. Much of the overshoot has been traced to an inhibitory mechanism in the retina such that the visual response from areas just outside of a white boundary is strongly repressed. A second source of the overshoot can be assigned to the combination of the small oscillatory motions of the eye and its variable-gain mechanism. For example, in viewing a black–white boundary, the visual gain is reduced in the white area and increased in the dark area. Now, if the eye shifts slightly so that the white image falls on a previously dark part of the retina, the visual response is enhanced. Similarly, a shift in the other direction yields a blacker than black response for the dark side of the boundary. A common example of these overshoots is the observation that the visual response to a photographic step-wedge is a sawtooth rather than a staircase.

The strong overshoots at sharp boundaries have the effect of reducing the ability of the eye to detect small differences in brightness near the boundary. Hence, the visibility of noise is significantly attenuated. A striking example of this effect is to view a noisy television raster from a distance and then to approach the raster. If the scanning lines are sharply defined so that there are dark spaces between them, the noise disappears at close viewing distances where the lines can be clearly resolved. The striking aspect, of course, is the anti-intuitive experience of finding that a fine texture, namely, the noise, that can be seen at large distances disappears at close distances.

5.5. Noise in Dark Areas

Both for visual images and for photographic film, the noise approaches zero as the signal approaches zero. This is as it should be for a photon-noise-limited picture. The same performance can be expected from those camera tubes in which some form of image multiplication ensures that the noise in the pattern of stored charges on the target dominates the noise in both the scanning beam and in the television amplifier. The image isocon approaches this performance to the extent that its noise is about a factor of 3 lower in the dark areas than in the lighted areas. In the vidicon and image orthicon, the noise remains substantially the same in the dark as in the lighted areas. In the vidicon it is the fixed noise added by the amplifier. In the image orthicon, it is the relatively fixed shot noise of the scanning beam.

The presence of a fixed noise can represent a significant loss in sensitivity depending, of course, on the picture content. If the emphasis is on the content in the high lights, no significant loss in sensitivity is incurred. If the emphasis is on the content in the low lights, then the incident light intensity must be increased in order to match the signal-to-noise ratio obtained in those devices in which the noise approaches zero at low lights. Hence, while the presence of a fixed noise does have its effect on sensitivity, it is not meaningful to attach a single number to its magnitude.

5.6. Noise *versus* Brightness of Reproduced Pictures

In Section 5.2, we pointed out the highly effective method for noise reduction consisting of interposing a neutral filter between the viewer and the picture on a television receiver. The inverse of this operation has had a significant effect on the quality of pictures demanded by the viewer.

In the course of the development of motion-picture film and as the technology of light sources improved, there has been a steady increase in the brightness of theater screens. The brightness has increased from a few foot-lamberts to almost 20 foot-lamberts. The increased screen brightness has necessitated a corresponding increase in film quality. At the higher light levels, the signal-to-noise ratio of the retinal image was enhanced to the point that the noise or graininess of the film was readily visible. Finer-grained films had to be used with a consequent increase in the light level or exposure required in the film studio.

Quantitatively, if the light flux entering the film camera from some element of the scene was the same as the light flux entering the viewer's eye from the same element, and if the quantum efficiencies of film and retina were equal, a noisefree picture would be seen in the theater. Further, if the lens diameters and viewing distances of camera and viewer were matched, the screen brightness would then be equal to that in the film studio. A higher screen brightness would then reveal a noisy picture. Actually, the brightness in film studios is some 10 or more times that of the theater screen. Part of this increased brightness reflects the several-fold-higher quantum efficiency of the eye as compared with photo-

a

Fig. 5.4. High-definition television picture (O.H. Schade). a) Full target scan. b) An underscanned portion of the television camera tube target. (See text for details.)

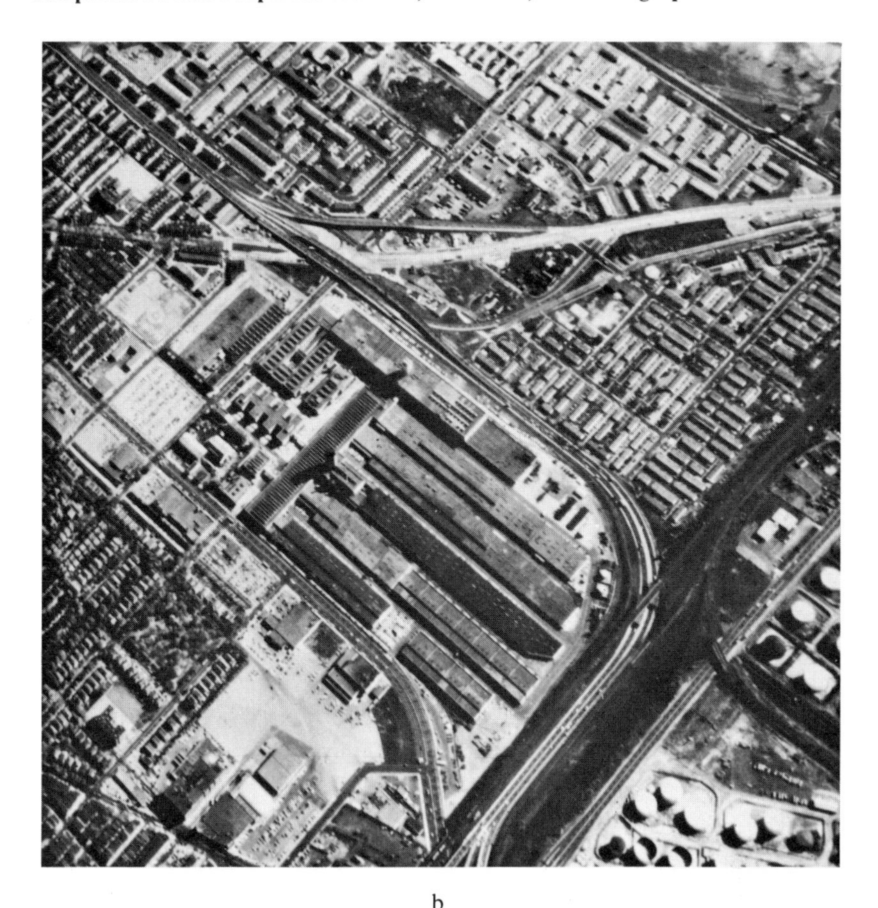

b

graphic film. Another and significant part has to do with the increased depth of focus required of the film camera as compared with the human eye. The human eye does not have a particularly large depth of focus. It compensates for this lack by rapidly refocusing on near or far elements of our surroundings as our attention shifts. The result is that we feel that we have a large depth of focus because everything that we look at (or become attentive to) is in focus. Since there is no way in which the camera operator can anticipate which part of the scene our attention will fasten on, all parts of the picture must be in good focus simultaneously, and available on demand. This necessitates a large depth of focus and a consequent increase in studio brightness relative to that of the theater screen.

The brightness of television screens is usually several times higher than that of theater screens and therefore requires a picture of higher quality in order to appear noisefree. This is reflected, for example, in the comparative highlight signal-to-noise ratio of a modern studio television picture and that of plus X, 35-mm film, both evaluated for a picture element size that is 1/500 of the picture size. For television, the signal-to-noise ratio is commonly 200, for plus X film it is about 80.[J-1] One would then expect to find films that are just noisefree on a theater screen to be detectably noisy on a television screen.

Parenthetically, the suggestion that television studio pictures might be of higher quality than motion pictures is almost invariably met with complete disbelief. The reason is that the comparison is made with television pictures as received in the home. For a variety of reasons, these pictures are likely to be significantly inferior to the quality of picture that the modern 500-line television system is capable of presenting. Moreover, this judgement is made by an observer who is several times more critical in judging the television picture than the motion picture owing to the four-fold increase in picture brightness.

Figure 5.4 is clear evidence that the picture quality transmitted by a high-definition television system can far exceed the quality of 35-mm motion-picture films as normally encountered. The original of Fig. 5.4a was a 9×9 inch photographic print. Figure 5.4a is then the image of the 9×9 inch print as transmitted by an 1800-line television system and displayed on the kinescope. The detail in Fig. 5.4a is still limited by the 1800 scanning lines. In order to show that the vidicon camera tube is capable of still greater resolution, the scanning pattern on the camera tube was reduced so that it reproduced only a small part of Fig. 5.4a. The resultant picture, Fig. 5.4b, confirms the 10,000-line capability of the camera tube. These pictures were transmitted and photographed by O.H. Schade on a system of his own design. Further details are to be found in his publication.[S-1]

Another aspect of the dependence of noisiness on the brightness of the reproduced picture is that a television system frequently finds itself acting in the role of a light amplifier. Pictures may be recorded in the fading afternoon light at a brightness level of a few foot-lamberts and displayed on the television screen at a brightness of some 80 foot-lamberts. In order to avoid presenting a noisy picture, the sensitivity of the television camera should exceed that of the human eye by the ratio of the brightness of the television screen to that of the original scene. Alternatively, the television camera can compensate for its lack of sufficient sensitivity by opening up its lens to admit more light. This operation, however, sacrifices depth of focus as compared with what a human observer would have at the scene. The recent super 8-mm, high-speed, color home motion picture cameras have moved strongly in this direction in order to record indoor pictures without extra illumination.

5.7. Summary

A proper comparison of signal-to-noise ratios of television, photographic, and human visual systems must be made for the same exposure times and for equivalent areas of picture elements.

The signal-to-noise ratio in a picture is, in general, proportional to the diameter of the picture element.

The noise-limited resolution of a picture is proportional to the contrast of the test pattern. The resolution appropriate to a contrast of 10% is only 1/10 that appropriate to a contrast of 100%.

A test pattern consisting of isolated discs of varying size and contrast is a more valid measure of resolution than the usual high-contrast bar patterns.

Higher presentation brightnesses require pictures of higher signal-to-noise ratios. Conversely, the most effective noise filter is a neutral filter interposed at the eye of the observer.

5.8. References

J-1. R. Clark Jones, "Quantum Efficiency of Detectors for Visible and Infrared Radiation," in *Advances in Electronics and Electron Physics*, Vol. 11, p. 147 (1959), Academic Press, New York.

M-1. R.H. Morgan, Threshold visual perception and its relationship to photon fluctuation and sine-wave response, *Am. J. Roentgenol. Radium Therapy Nucl. Med.* **93**, 982–996 (1965).

S-1. O.H. Schade, Electron optics and signal read-out of high definition return-beam vidicon cameras, *RCA Rev.* **31**, 60–119 (1972).

General

A. Rose, A unified approach to the performance of photographic film, television pickup tubes and the human eye," *J. Soc. Motion Picture Engrs.* **47**, 273–294 (1946).

CHAPTER 6

IMAGE MULTIPLIERS

6.1. Introduction

This chapter is necessarily brief, not as a measure of its importance, but rather as a measure of the simplicity of the principles involved and the relatively high degree of perfection of the available devices.

The ultimate goal of any detector of light is the counting of individual photons. Photoemitters combined with photomultipliers were able to achieve this goal for single-element devices already in the 1930's. Image tubes were able, in principle, to do the same at a comparably early stage. In practice, the technology of image tubes required some years of development before the highly sensitive, multistage image multipliers of the last decade could be realized. The technology included the improvement of the quantum efficiency of photocathodes,[(S-1)] the improvement of the luminescence efficiency of phosphors, the reduction in thermally generated dark current, the development of compact high-voltage sources, and the stage-to-stage coupling via fiber-optic face plates.

The progress of vacuum-tube image multipliers was early and rapid. The same multiplication which can in principle be far more compactly achieved in a solid has been painfully slow in realization. The difference, of course, is that a photoemitted electron can be accelerated to hundreds or thousands of volts by the mere application of these voltages. The mean free path of an electron in vacuum is the separation of cathode and anode, and can be indefinitely large. The mean free path of a photoexcited electron in a solid is likely to be only a few angstroms, in which distance it acquires only a few hundredths of a volt. It is exceedingly difficult, as will be discussed in Chapter 9, to achieve even an energy of the few volts required for impact ionization in relatively insulating materials without an attendant and objectionable increase in dark current via field emission.

It is interesting also to contrast the electrical energy of the photon, which first excites an electron, with the electrical energy of the liberated electron when it is used on the control electrode of some triode amplifier. The photon energy in the visible range is some 2 electron volts. It is a hundred times larger than the room-temperature thermal energy. Hence, the probability of a competing thermal excitation of 2 volts is vanishingly small. The number of times per second that an atom or molecule will achieve 2 volts by thermal excursions is

$$\nu \exp(-80) \sim 10^{-21}/\text{sec}$$

where the largest value ν is the frequency of lattice vibrations, namely, 10^{14}. For this reason, photoexcitations in the visible range are almost totally free from competition by thermal processes.

Once an electron is liberated from its bound state and deposited on the control electrode of a triode amplifier, the signal voltage of the electron plummets from its initial value of 2 volts to a value of microvolts or less depending on the capacitance of the control electrode. Its value is

$$V = e/C = 10^{-8}\ \text{volt} \qquad \text{for} \quad C = 10^{-11}\ \text{farad}$$

Under these conditions the signal voltage is obscured by the noisy thermal voltages of the resistor tied to the control electrode. If we wished to match the 2 volts of the original bound electron, the dimensions of the control grid would need to be about 1 Å (10^{-8} cm), or about the same atomic dimension of the original bound electron.

We will return to this point in the next chapter on solid-state photon counters. For the present, we wish only to call attention to the dramatic degradation of energy that accompanies the liberation of a bound electron and its deposition on a control electrode.

6.2. Varieties of Image Multipliers

Figures 6.1 to 6.8 sketch the various forms that image multipliers have taken. They are mostly self-explanatory. The structure in Fig. 6.1 is the only one in which an opaque photocathode is used.(C-1) The successive images suffer a shear distortion peculiar to the use of cylindrical electric and magnetic fields for focusing.(R-1) The shear distortion is shown for a single stage in Fig. 6.2.

Figure 6.3 shows a simple concatenation of image tubes.(M-1) The drawback of this arrangement is the large loss of light between stages since the lens collects only a small fraction of the light from each fluorescent screen.

In Fig. 6.4, the electron image is multiplied by a succession of fine mesh screens. The electron paths are collimated by an axial magnetic field. This arrangement suffers from a small gain per stage and from the defocusing that accompanies the large spread in energies (several volts) of the secondary electrons.

In Fig. 6.5, an array of fine tubes or channels is used to mechanically collimate the electrons.(W-2) The channels are lined with a resistive coating so that several thousand volts can be applied from their entrance ends to their exit ends. Considerable effort has been expended on the technology of forming, coating, and sensitizing these channels. The arrangement is remarkably compact. It suffers from an excess of noise owing to the uncertain nature of the multiplication process. That is, the number of stages of multiplication that each electron suffers is a sensitive function of its emission velocity and its direction of emission.

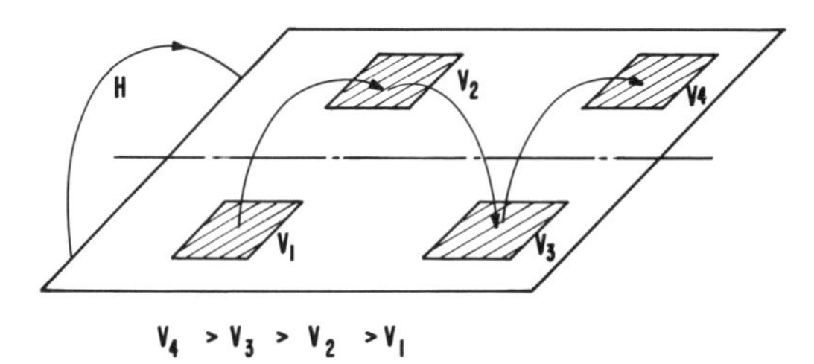

Fig. 6.1. Image multiplier using cylindrical electric and magnetic fields.

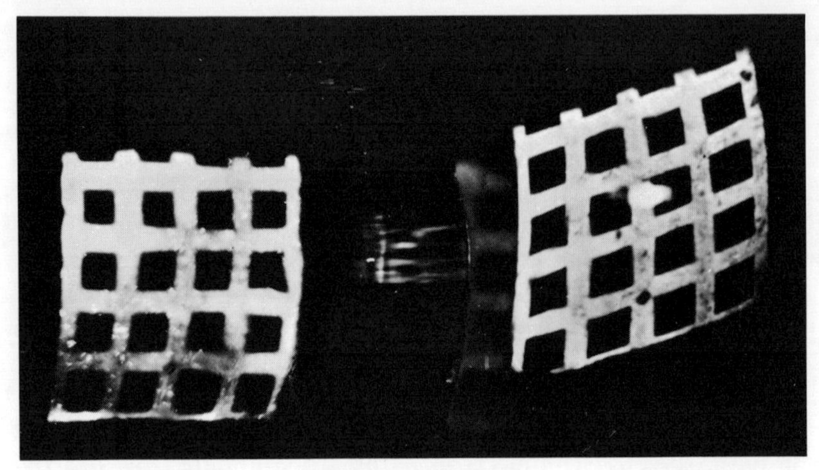

Fig. 6.2. Illustration of shear distortion characteristic of imaging in cylindrical electric and magnetic fields. The optical image on the cathode is shown on the left; the electron image on the anode is on the right.

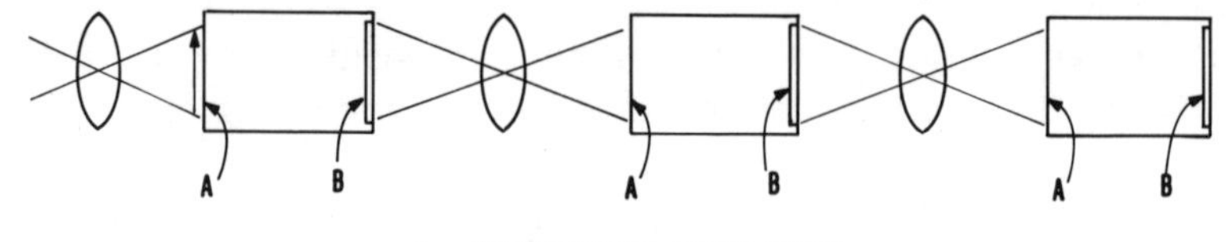

Fig. 6.3. Single-stage image intensifiers coupled by lenses.

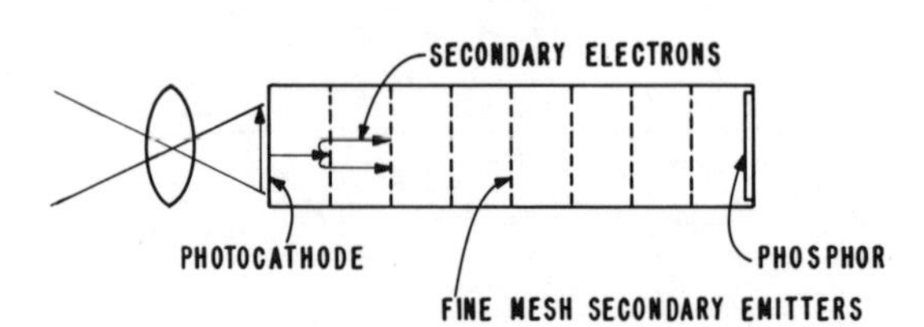

Fig. 6.4. Image intensifier using a series of secondary-emission screens.

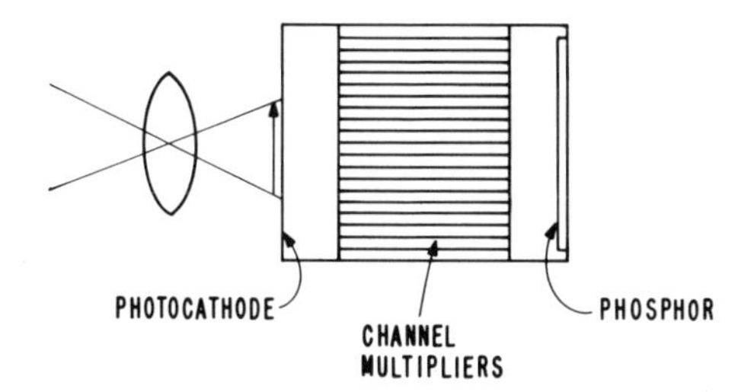

Fig. 6.5. Channel-plate image intensifier.

In Fig. 6.6, very thin sheets of material,[W-1] only microns thick, are used as transmission secondary emitters. The primary electrons incident on one face excite hundreds of secondary electrons in the volume of the secondary-emitting film. Only a few of these manage to escape the far side. The energy needed to escape is extremely rapidly dissipated by emission of phonons in the solid. As will be shown in Chapter 9, the rate of energy emission to phonons is of the order of 10^{13} eV/sec or 10^{6} eV/cm of random path. This form of image multiplier may very well experience a revival if the secondary-emission ratio of transmission secondary emitters can be improved using the modern technology of negative-affinity emitters. Here, even when the energetic, volume-excited secondary electron settles down to the conduction band, it can still be emitted since the conduction band lies above vacuum (see Fig. 6.7).

Figure 6.8 shows the more generally used combination of phosphor and photoemitter tightly coupled at each stage by

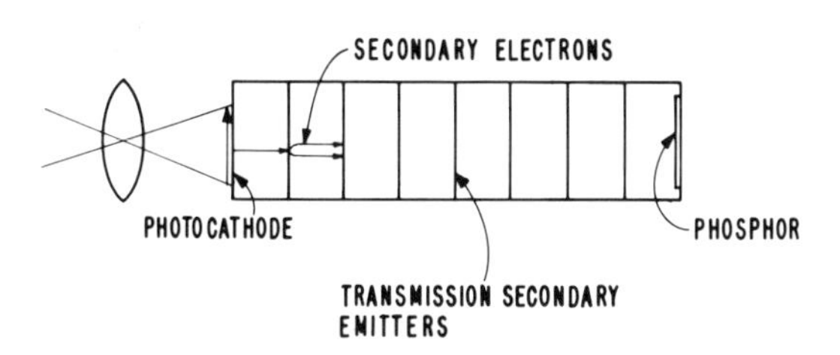

Fig. 6.6. Image intensifier using a series of transmission secondary-emission films.

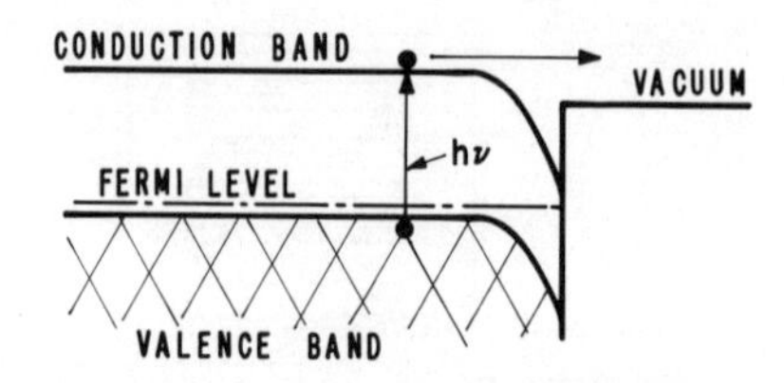

Fig. 6.7. Negative-affinity photoemitter.

being located on opposite faces of a thin sheet of mica or other insulator.[R-2] The gain per stage is in the range of 50–100 with about 10^4 volts applied per stage. The gain can be estimated from the combination of a phosphor energy efficiency of 10%, and a photoemitter yield of 10%. The overall efficiency of 1%, or 10^{-2}, means that 100 volts out of the applied 10^4 volts emerge as the energy of photoemitted electrons. Since photoemission in the visible requires about 2 volts per emitted electron, each stage yields some 50 emitted electrons per incident primary electron. Note that in this arrangement the same 10^4 volts applied across the first stage can be used for the second and third stages, providing the thin insulator sheets can support 10^4 volts without breakdown.

Figure 6.9 is essentially a duplicate of Fig. 6.8 except that the coupling between phosphor and photoemitter is accomplished by fiber-optic face plates. These are bundles of fine glass fibers fused into a single block. Light entering one end of a glass fiber is confined to that fiber by total internal reflection and emerges at the other end with negligible attenuation. The use of fiber-optic couplers

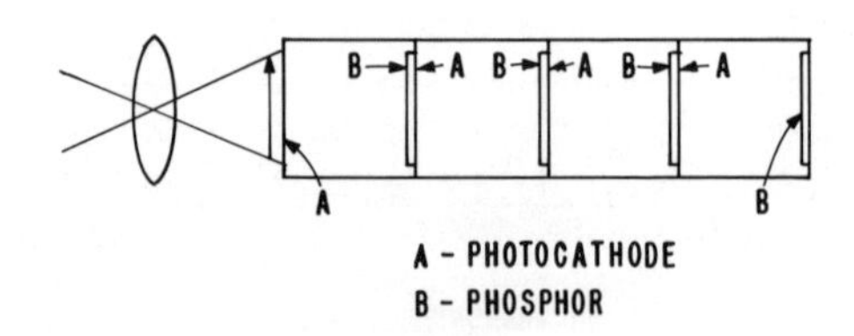

Fig. 6.8. Image intensifier using direct coupled phosphor–photocathode elements.

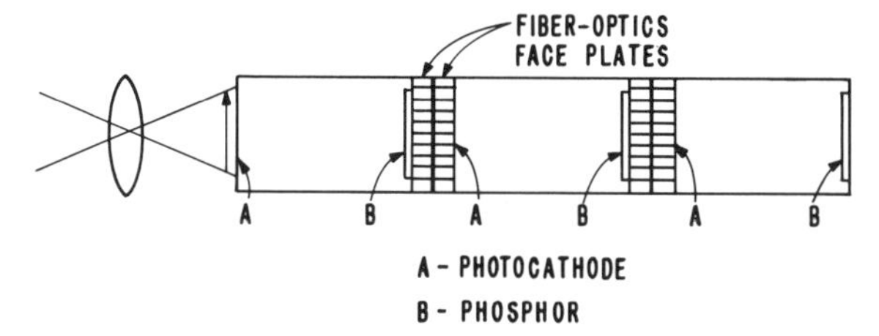

Fig. 6.9. Image intensifiers coupled by fiber-optics face plates.

allows the several stages to be manufactured separately and, thereby, contributes to an improved manufacturing yield. The fiber optics also permit the use of curved photocathode surfaces so that electrostatic focusing of the image can be achieved with negligible distortion.

Figure 6.10 is a representative image from the output of a three-stage image multiplier when the image tube was focused on a night scene. The lens opening was $f/1.4$ and the exposure time was 1/250 sec.

Fig. 6.10. Representative image of a three-stage image multiplier focused on a night scene (courtesy of Richard D. Faulkner). The exposure time was 1/250 sec.

6.3. Typical Performance of a Three-Stage Image Multiplier

The following data are characteristic of recent RCA three-stage image multiplier tubes. The tubes are about 8 inches long and 3 or 4 inches in diameter. The focusing is via a uniform magnetic field of a few hundred gauss giving a 1:1 magnification ratio. The overall gain is in the order of 5×10^4 photons out per photon in. The input or first-stage photocathode has a quantum efficiency of about 10% in the visible range.* The gain per stage is sufficiently high (~ 40) to ensure that negligible noise is introduced by the multiplication process. The resolution is about 400 television lines/cm, or some 10^5 picture elements/cm^2.

Finally, the dark current is of interest in setting the lowest value of input light intensity. The dark current is in the order of 10^5 electrons/cm$^2 \cdot$ sec from the first photocathode. This is the equivalent of a bias light of 10^6 photons/cm$^2 \cdot$ sec, or 10^{-7} foot-candle on the first photocathode. It is also of the same order as the illumination of the human retina at its absolute threshold. The 10^5 electrons/cm$^2 \cdot$ sec also correspond to 0.1 electrons per high-light-resolvable picture element in the visual storage time of 0.1 sec. Alternatively, the 10^5 electrons/cm$^2 \cdot$ sec correspond to a resolution of some 30 lines/cm for the 0.1 sec storage time and for a 100% contrast in the test pattern. Lower-contrast test patterns would yield a proportionately lower number of lines/cm.

The dark current can be markedly reduced by a modest cooling, even to dry ice temperature.

The gain of the three-stage image multiplier is still too low to see the trace of individual photons with the naked eye. They should, however, be visible with a small magnifying glass providing the phosphor decay time is less than a tenth of a second.

In summary, the image multiplier is a photon-noise-limited device that operates with a quantum efficiency of 10% down to an incident brightness of 10^{-8} foot-candle.

*Peak quantum efficiencies of modern photocathode reach as high as 40%.

6.4. Summary

Image multipliers have achieved a high state of technological development and performance.

A three-stage image multiplier provides a light amplification approaching a millionfold.

Photon-noise-limited performance with a quantum efficiency of 10% can be achieved down to scene brightnesses of about 10^{-8} foot-lambert.

6.5. References

C-1. F. Coeterier and M.C. Teves, Apparatus for the transformation of light of long wavelength into light of short wavelength, *Physica* **3**, 968–976 (1936).

M-1. G.A. Morton, Image intensifiers and the scotoscope, *Appl. Opt.* **3**, 651–672 (1964). In this review of image intensifiers, Morton cites early papers on the image tube by: G. Holst, J.H. de Boer, M.C. Teves, and C.F. Veenamans, *Physica* **1**, 297 (1934); E. Bruche and W. Schaffernicht, *Elek. Nachr.-Tech.* **12**, 381 (1935); V.K. Zworykin and G.A. Morton *J. Opt. Soc. Am.* **26**, 181 (1936); W. Heiman, *Elek. Nachr.-Tech.* **12**, 68 (1935).

R-1. A. Rose, Electron optics of cylindrical electric and magnetic fields, *Proc. IRE.* **28**, 30–40 (1940).

R-2. Morton in Ref. (M-1) cites early work on this arrangement by J.E. Ruedy in 1941.

S-1. A.H. Sommer, *Photoemissive Materials Preparation, Properties, and Uses,* John Wiley & Sons, Inc., New York (1968).

W-2. W.C. Wiley and C.F. Hendee, Electron multipliers utilizing continuous strip surfaces, *IRE Trans. Nucl. Sci.* **NS-9**, 103 (1962).

W-1. W.L. Wilcock, D.L. Emberson, and B. Weekley, Work at Imperial College on image intensifiers with transmitted secondary electron multiplication, *IRE Trans. Nucl. Sci.* **NS-7**, 126 (1960).

General

L.N. Biberman and S. Nudelman, *Photoelectronic Imaging Devices,* Vols. I and II (1971), Plenum Press, New York.

G.A. Morton, Image intensifiers and the scotoscope, *Appl. Opt.* **3**, 651–672 (1964).

H.V. Soule, *Electro-Optical Photography at Low Illumination Levels* (1968), John Wiley & Sons, Inc., New York.

"Symposium on Photoelectronic Image Devices," in *Advances in Electronics and Electron Physics,* Vols. 12, 16, and 24A, B (1960, 1962, and 1966), Academic Press, New York.

CHAPTER 7

SOLID-STATE PHOTON COUNTERS

7.1. Introduction

The counting of photons by photoemission into vacuum and subsequent electron multiplication has long been an accomplished fact. The solid-state version of the photomultiplier, using high-field impact ionization by photoexcited carriers, has been extremely difficult to achieve. Only some fringe evidence of its realizability has been published in the last few years. This problem will be discussed in more detail in Chapter 9.

The present chapter is concerned with the ability of low-voltage devices such as photoconductors and transistors to count individual photons. We will analyze the performance of single-element detectors. The same detectors can, however, be arranged in an array to form an image sensor or be used as a single-element detector in combination with a scanning system which scans the image past the element. Examples of the latter are the mechanical scanning of an image across a photoconductor, the scanning of the scene by a flying light spot, and the recent self-scanned image sensors which use the bucket-brigade or charge-coupled principle to feed the video signal into an MOS with a floating gate.

Good evidence for "solid-state" photon counters was cited in the chapters on the visual process and the photographic process. In the visual process, the "solid-state" material is the complex biological material rhodopsin. The mechanism for amplifying the energy of a single photon to the energy, over a millionfold higher, of a nerve pulse, while necessarily of some generalized catalytic form, has yet to be delineated. In the photographic process, it is clear that the deposition of a few silver atoms by the absorption of a few photons in silver bromide is sufficient to trigger off the chemical development of an entire silver bromide grain into some 10^9 silver atoms. In both cases, visual and photographic, the primary photoexcitations are protected against competition from thermal processes by the magnitude of the energy, some 2 volts, required for excitation. After excitation, a new species is created, either a conformal change in rhodopsin or a silver atom, which did not exist before. Hence, it is easy to understand, in a formal sense, why these photoexcitations can yield extremely high gains and stand out clearly against a thermal background.

We contrast the biological and photographic processes with the purely electronic processes required to count photons by a photoconductor or solid-state triode. Here, again, the primary photoexcitation is protected against thermal competition by virtue of the 2-volt excitation energy. But, once the electron is excited, it must give rise to a distinguishable pulse of current or charge—distinguishable against a background of pre-existing electrons whose presence is essential to the workings of the photoconductor or triode. The pre-existing free electrons generate large direct currents and smaller, fluctuating, thermally induced noise currents whose magnitudes are normally large compared with the signal current generated by the single additional photoexcited electron. The problem, then, of detecting one additional photoexcited electron against a pre-existing sea of thermally generated electrons is a delicate one. It is sufficiently delicate that the technology of modern solid-state materials is only beginning to yield the required degree of purity and control.

The problem of detection of single photons is closely linked with the nature and magnitude of noise currents in solids. For this reason the discussion begins with a treatment of noise currents. The treatment[R-1] is somewhat unconventional and emphasizes the particle, as opposed to the Fourier, aspects of noise so that the particle noise limitations can be readily compared with the particle detection capabilities.

7.2. Noise Currents and Charges

7.2.1. Thermal Noise

Consider one of the electrons in the conduction band of a semiconductor (Fig. 7.1). In a time t it will make t/τ random collisions with the lattice. τ is the time for a single collision ($\approx 10^{-13}$ sec). On the average, the electron will remain at its starting point since, on the average, it makes an equal number of excursions in the positive and negative directions. However, since these are random collisions, there will be an rms deviation from the average behavior, measured by the square root of the number of collisions, that is, by $(t/\tau)^{1/2}$. The rms departure of the electron from its starting point is then

$$d = \pm \left(\frac{t}{\tau}\right)^{1/2} l \tag{7.1}$$

where l is the mean free path per collision. The charge contributed to the outside circuit is

$$q_n = \pm \frac{ed}{L} = \pm e \left(\frac{t}{\tau}\right)^{1/2} \frac{l}{L} \tag{7.2}$$

where L is the electrode spacing. The rms fluctuation in charge Q_n contributed to the outside circuit by the total number N of electrons is $N^{1/2}q_n$ since the individual contributions (like the standard problem of coin tossing) are randomly positive and negative. Hence,

$$Q_n = N^{1/2}q_n = \pm e \left(\frac{Nt}{\tau}\right)^{1/2} \frac{l}{L} \tag{7.3}$$

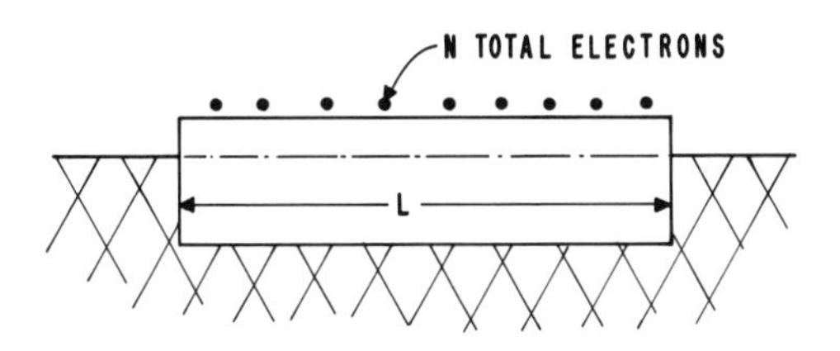

Fig. 7.1. Model semiconductor used for analysis of noise currents.

The largest value of time t during which the noise charge can be accumulated is given by the dielectric relaxation time, RC, of the semiconductor. This is the characteristic time for relaxing a charge or electrostatic field by ohmic conduction. Hence, the largest value of Q_n is

$$Q_n = e\left(\frac{NRC}{\tau}\right)^{1/2}\frac{l}{L} \tag{7.4}$$

We make use of the following relations,

$$\text{Resistance} \equiv R = \frac{L^2}{Ne\mu}$$

$$\text{Mobility} \equiv \mu = \tau\frac{e}{m}$$

$$\text{Thermal energy} = kT = \tfrac{1}{2}mv^2 = \tfrac{1}{2}m\frac{l^2}{\tau^2}$$

Capacitance of semiconductor $\equiv C$
(or total capacitance across ends of semiconductor)

to convert Eq. (7.4) into

$$Q_n = (2kTC)^{1/2} \tag{7.5}$$

We note several significant features of Eq. (7.5). First, it is independent of the resistance of the semiconductor sample. Second, the capacitance C can be increased by adding an extra lumped capacitance across the ends of the semiconductor. And third, Eq. (7.5) yields the well-known expression for thermal noise currents I_n,

$$I_n \equiv \frac{Q_n}{RC} = \left(\frac{2kT}{R^2C}\right)^{1/2} = \left(\frac{4kT}{R}\Delta f\right)^{1/2} \tag{7.6}$$

where we have taken the band with Δf to be

$$\Delta f \doteq \frac{1}{2RC}$$

We will make use of Eq. (7.5), giving the thermally induced noise charge, and also make use of a second relation for the noise charge resulting from a random current of particles, the so-called shot noise. This will be derived next.

7.2.2. Shot Noise

Consider two boxes A and B, as in Fig. 7.2, containing N_A and N_B particles, respectively, in the steady state. Let there be a stochastic exchange current I of particles between the boxes. The current I has a constant average value but, like photoemission, it is subject to the usual statistical fluctuations. The current *from* each box is understood also to be proportional to the number of particles in that box. The proportionality constants are, of course, inversely proportional to the numbers N_A and N_B. In the time t, the number of particles exchanged is

$$N_t = 2It$$

The rms fluctuation in this number exchanged is then

$$\delta N = (N_t)^{1/2} = (2It)^{1/2} \tag{7.7}$$

The current I can be written formally as

$$I = \frac{N_A}{\tau_A} = \frac{N_B}{\tau_B} \tag{7.8}$$

where τ_A and τ_B are the average lifetimes of particles in their respective boxes. Hence

$$\delta N = \left(2\frac{N_A}{\tau_A}t \right)^{1/2} = \left(2\frac{N_B}{\tau_B}t \right)^{1/2} \tag{7.9}$$

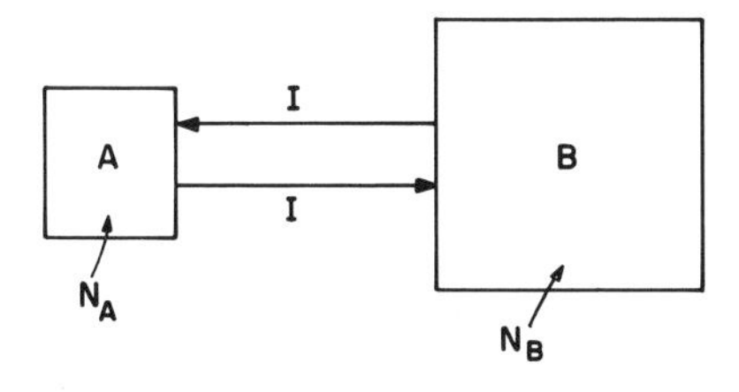

Fig. 7.2. Model used for analysis of shot noise.

Now we ask, what is the largest time t over which particle fluctuations can be accumulated in each box? The answer is that this time is τ_A, namely, the shorter lifetime. This follows from the definition of lifetime. It is the time required for a small departure from the average number N_A (or N_B) to be relaxed via a proportionately small departure from the average current I. Since these departures are damped out faster by the smaller lifetime and since, by conservation of particles, the magnitude of fluctuations must be the same in both boxes, the maximum fluctuation δN is

$$\delta N = \left(2 \frac{N_A}{\tau_A} \tau_A \right)^{1/2} = (2N_A)^{1/2} \tag{7.10}$$

that is, *the square root of the smaller number of particles.* If these particles are electrons, the rms fluctuation in charge is

$$Q_n = e\delta N = e(2N_A)^{1/2} \tag{7.11}$$

Equation (7.11) immediately yields the conventional expression for shot noise by introducing the characteristic relaxation time

$$\tau_A = \frac{1}{2\Delta f}$$

and computing the noise current for one of the two exchange currents.

$$I_n = \frac{Q_n}{\tau_A} = \left(2e \frac{eN_A}{\tau_A} \frac{1}{2\tau_A} \right)^{1/2}$$
$$= (2eI\Delta f)^{1/2}$$

Note that if, in Fig. 7.2, a third box C were exchanging currents with B, and if N_C also were much less than N_B, the noise charge in B would be

$$Q_n = e(2N_A + N_C)^{1/2} \tag{7.12}$$

That is, A and C and other boxes make independent noise contributions to B providing the number of particles in B is large compared with the sum of the particles in the other boxes. The lifetime of particles in B must be sufficiently large that, for example, fluctuations in B due to A are not relaxed by changes in current from

B to *C*. Another way of stating this condition is that

$$\tau_B \gg \tau_A + \tau_C + \tau_D + \dots \tag{7.13}$$

in order to insure independent noise contributions from the sources *A*, *C*, *D*, etc. to *B*. These considerations are important to our analysis, below, of the masking effect of $1/f$ noise currents on the detectability of single photons.

7.3. Photoconductive Photon Counter

7.3.1. Trapfree Photoconductors

The problem is to outline the conditions under which a photoconductor at the input of a conventional amplifier can detect the excitation of a single free electron by a single photon.[R-2] The photoconductor has a finite dark resistance which we take to be the input resistance of the amplifier. Normally, the amplifier can see the thermal noise of its input resistance. Hence, we only have to derive the conditions under which the photocurrent due to a single photoexcited electron will exceed the thermal noise current. Actually, it is simpler to examine the problem in terms of noise charges accumulated at the input electrode in the relaxation time of the input circuit.

In Fig. 7.3 we show the significant parameters of the input circuit. The input resistance is taken equal to that of the photoconductor. Larger values would have no effect since they would be shunted by the photoconductor. Smaller values would reduce the signal voltage due to photocurrents faster than the noise voltage of the lower resistance. The input capacitance C_A is taken

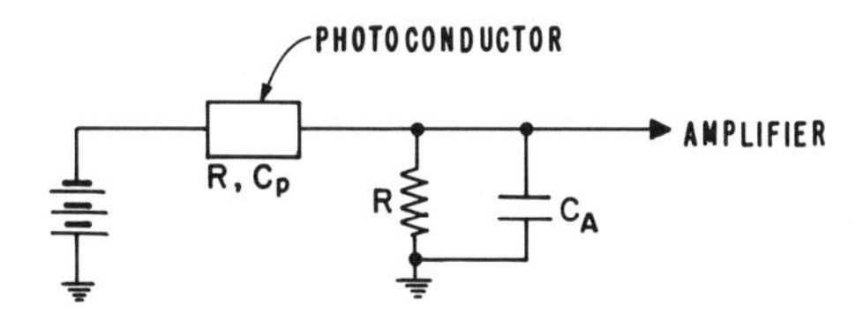

Fig. 7.3. Combination of photoconductor and amplifier for photon counting.

to be large compared with that of the photoconductor for reasons that will become evident shortly. The input time constant of the amplifier is then

$$\tau = RC_A \tag{7.14}$$

By Eq. (7.5) the input noise charge is

$$Q_n = (2kTC_A)^{1/2} \tag{7.15}$$

We need only find the conditions under which the signal charge due to one photon exceeds the noise charge.

An energy band diagram of the photoconductor is sketched in Fig. 7.4. We show the usual ohmic contacts, a certain density of electrons already present in the dark, and a density of states, lying well below the Fermi level, out of which the photon excites a free electron. We have omitted the shallow trapping states normally present in every semiconductor. These will be introduced following the present argument. We further assume that the lifetime of a photoexcited electron against recombination can be freely adjusted and we take it equal to the RC_A time of the amplifier. Larger values would yield no more photocharge in the time $\tau = RC_A$, and smaller values would yield less.

The signal charge due to one photoexcited electron is then

$$Q_s = e\frac{\tau}{T_r} \tag{7.16}$$

where T_r is the transit time of a free electron across the photoconductor. Since the transit time can be decreased by increasing the field across the photoconductor, one might think that there

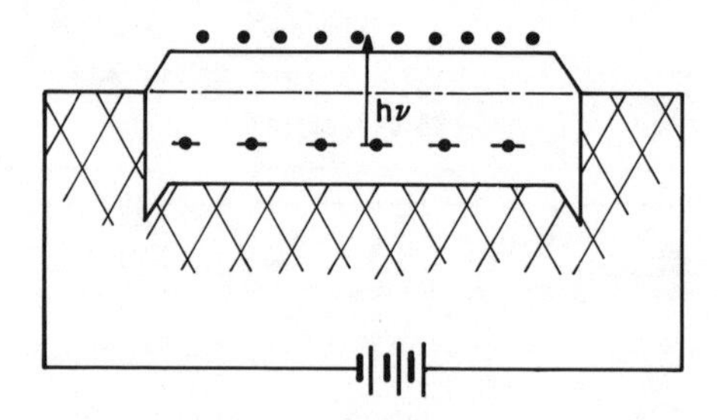

Fig. 7.4. Ideal photoconductor for photon counting.

is no constraint to making the signal charge Q_s exceed the noise charge Q_n. The fact is that there is an upper limit to the field that can be applied without increasing the conductivity of the photoconductor in the dark. This is the phenomenon of space-charge-limited currents in solids. When the charge on the anode exceeds the total charge of the free carriers in Fig. 7.4, the anode injects additional free charge into the volume of the photoconductor to match its own charge. Hence the dark conductivity increases.

The dark conductivity, then, sets the upper limit of the field that can be applied across the photoconductor. We can increase this field by choosing a higher dark conductivity. This would, of course, result in a smaller value for τ.

A brief argument shows that the field at the onset of space-charge-limited currents yields a transit time equal to the RC_P product for the photoconductor. The condition for space-charge-limited currents is

$$K\mathscr{E} = 4\pi neL \tag{7.17}$$

where $\mathscr{E}$ is the field, K is the dielectric constant, and n is the density of free electrons. We multiply both sides by the mobility μ and rearrange factors to obtain

$$\frac{K}{4\pi ne\mu} = \frac{L}{\mathscr{E}\mu} \tag{7.18}$$

The left-hand side is the RC_P product of the photoconductor; the right-hand side is the transit time for a free electron.

Hence the signal charge of Eq. (7.16) can be written

$$Q_s = e\frac{\tau}{T_r} = e\frac{\tau}{RC_P} = e\frac{C_A}{C_P} = eG \tag{7.19}$$

where G is the photoconductive gain of the photoconductor. We now equate this charge to the noise charge of Eq. (7.15) to obtain the lower limit of gain required for the photoconductor

$$G \geqq \frac{1}{e}(2kTC_A)^{1/2}$$

or

$$G \geqq \frac{2kTC_P}{e^2} \tag{7.20}$$

Alternatively, we can solve for the lower limit of conductivity σ for the photoconductor

$$eG = e\frac{\tau}{RC_P} \doteq 10^{12}\, e\frac{\tau 4\pi\sigma}{K} \geqq (2kTC_A)^{1/2}$$

or

$$\sigma \geqq 10^{-12}\frac{K(2kTC_A)^{1/2}}{4\pi e}\frac{1}{\tau} \qquad (7.21)$$

where $1/\tau$ is the fastest rate at which single photons can be separately detected.

A representative value for the gain G from Eq. (7.20) is 10^3, assuming room temperature and $C_A = 10^{-11}$ farads. Similarly, from Eq. (7.21),

$$\sigma \gtrsim \frac{10^{-9}}{\tau}(\text{ohm}\cdot\text{cm})^{-1}$$

Thus, for example, the conductivity of the photoconductor should be greater than 10^{-3} $(\text{ohm}\cdot\text{cm})^{-1}$ in order to count photons at the rate of 10^6/sec.

7.3.2. *Trapping Effects*

The introduction of shallow traps, those lying between the Fermi level and the conduction band, does not alter the argument carried out in the preceding section. The key expression was the signal charge passed through the photoconductor due to one photon at a voltage across the photoconductor sufficient for the onset of space-charge current flow. That is, in Eq. (7.16)

$$Q_s = \frac{\tau}{T_r}$$

where τ is the lifetime of a free electron. τ is also the response time of the photoconductor in the absence of traps. T_r is the transit time of a free electron at the onset of space-charge-limited current flow. The introduction of shallow traps causes the response time to exceed the lifetime by the relation

$$\tau_{\text{life}} = \frac{n_f}{n_t}\tau_{\text{resp}} \qquad (7.22)$$

where n_f and n_t are the densities of free and trapped electrons, respectively. The presence of traps also increases the voltage at which space-charge-currents set in, since the applied field must be sufficient to double the density of both the free electrons and trapped electrons with which they are in equilibrium. Hence, the transit time, which by Eq. (7.18) in the trapfree case was equal to the ohmic relaxation time RC_P of the photoconductor, is now given by

$$T_r = \frac{n_f}{n_t} RC_P \tag{7.23}$$

Insertion of Eqs. (7.22) and (7.23) into Eq. (7.16) yields

$$Q_S = e\frac{\tau_{\text{resp}}(n_f/n_t)}{RC_P(n_f/n_t)} = e\frac{\tau_{\text{res}}}{RC_P} = e\frac{C_A}{C_P} \tag{7.24}$$

namely, the same result as in Eq. (7.19) for the trapfree case. Hence, Eqs. (7.20) and (7.21) which define the gain and conductivity for a trapfree photoconductor are also valid for the photoconductor with shallow traps.

The discussion thus far has assumed that the only noise to be surmounted is the thermal noise of the resistance of the photoconductor. This would certainly be true for the trapfree photoconductor. On the other hand, the introduction of traps is likely to bring with it an excess noise associated with some kind of trapping process. In general, as the voltage across a photoconductor or semiconductor is increased, an excess noise is introduced whose mean squared noise current per unit bandwidth varies as f^{-1}. While this behavior is universal, the identification of the noise sources in a particular device has been difficult and subject to considerable disagreement. We will cite one likely source in the discussion of MOS triodes below. First, it will be useful to analyze the formal character of $1/f$ noise sources in such a way that their competition with the detection of single photons can be readily assessed.

7.4. An Analysis of $1/f$ Noise

Consider, as in Fig. 7.5, that the free electrons of a semiconductor are in equilibrium with a *smaller number* N of trapped

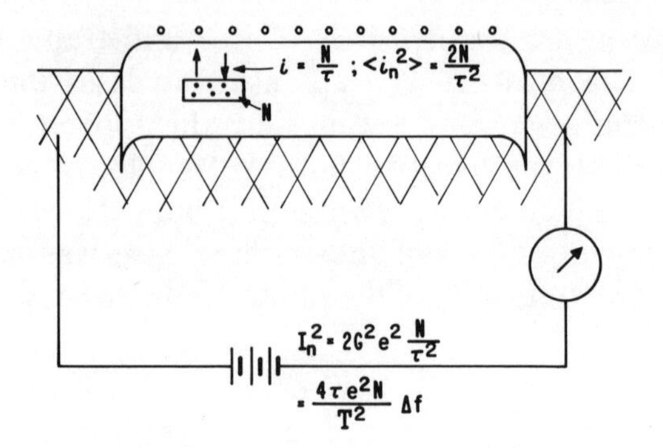

Fig. 7.5. Model used for analysis of excess noise arising from a single set of trapped electrons.

electrons whose lifetime in traps is τ. There will then be an exchange current

$$i = \frac{2N}{\tau} \tag{7.25}$$

between the free and trapped electrons. There will also be a corresponding noise component of this current (see Section 7.2)

$$i_n = \frac{(2N)^{1/2}}{\tau} \tag{7.26}$$

In the external measuring circuit, the observed noise current corresponding to Eq. (7.26) will be

$$I_n = \frac{Ge(2N)^{1/2}}{\tau} \tag{7.27}$$

since each electron generated from the trapped to the free states contributes a charge to the external circuit

$$Ge = e\frac{\tau}{T_r} = e\frac{\tau V\mu}{L^2} \tag{7.28}$$

during its lifetime τ. T_r is the free-electron transit time across the semiconductor. Note that the lifetime τ, for the reasons given in Section 7.2, is referred to the smaller group of electrons, namely, the trapped electrons.

The mean squared current of Eq. (7.27) can be written

$$I_n^2 = -\frac{4G^2e^2N}{\tau}\Delta f \tag{7.29}$$

where $\Delta f(= 1/2\tau)$ is the passband required to resolve the time interval τ. Or, using Eq. (7.28) for Ge, the spectral density of the noise current is

$$\frac{I_n^2}{\Delta f} = \frac{4V^2\mu^2e^2N\tau}{L^4} \tag{7.30}$$

and as shown in Fig. 7.6, is a flat spectrum cutting off at $f = 1/2\tau$.

In order to fabricate the $1/f$ noise spectrum, we add other trapped-electron noise sources as in Fig. 7.7. Here, each source has the same number of trapped electrons N, but their lifetimes in traps are τ, 2τ, 4τ, etc. Hence, each source contributes a mean-squared-noise amplitude a factor of 2 higher than the preceding source but extending only to a cutoff frequency a factor of 2 lower. The sum of the contributions of these noise sources is shown in Fig. 7.8 and is approximately described by Eq. (7.30) rewritten in the form

$$\frac{I_n^2}{\Delta f} = 4\frac{V^2\mu^2e^2}{L^4 f}N \tag{7.31}$$

where $f = 1/2\tau$. Equation (7.31) is the envelope of the sum of the contributions from the set of discrete sources. The set of discrete sources was chosen for ease of exposition and, as well, to define

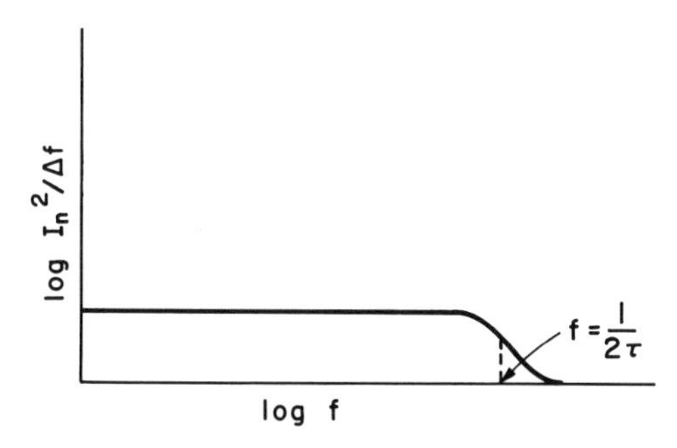

Fig. 7.6. Flat noise spectrum resulting from a single set of trapped electrons.

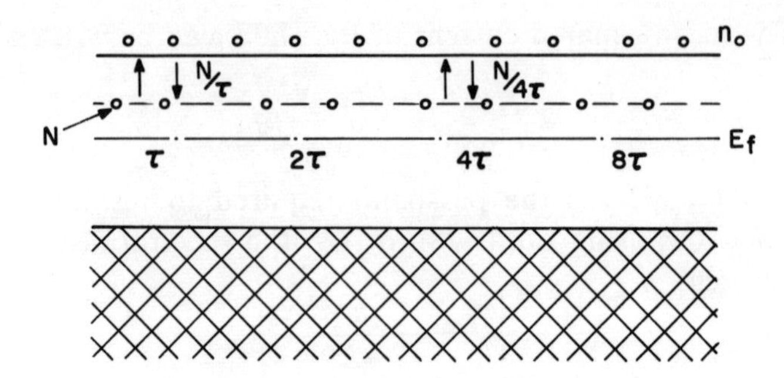

Fig. 7.7. Model for $1/f$ noise consisting of multiple sets of trapped electrons with different trapping times.

the key parameter N. A smoothed-out distribution of sources would satisfy Eq. (7.31) directly.

The parameter N is the number of trapped electrons per "time slot." The time slot is the range of lifetimes τ–2τ, 2τ–4τ, etc. In this way, one has a highly compact and graphic way of thinking about a variety of possible $1/f$ noise sources and of quantitatively

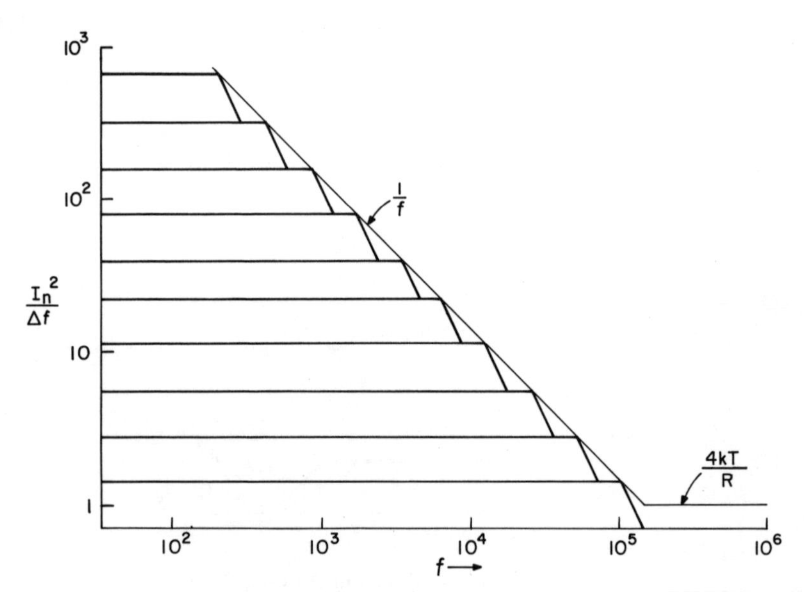

Fig. 7.8. $1/f$ noise spectrum constructed from noise sources in Fig. 7.7.

evaluating their noise contributions. Note that even when the source of $1/f$ noise is not trapped electrons, the observed spectrum can be analyzed by Eq. (7.31) to give the "equivalent number of trapped electrons."

An immediate example of the convenience of this designation is the fact that the fluctuation in the number of trapped electrons for a passband from f_1 to f_2 is (see Section 7.2)

$$\delta N = \left[N \log_2 \frac{f_2}{f_1} \right]^{1/2} \tag{7.32}$$

or the square root of the number of trapped electrons that lie in the frequency range f_1–f_2. By conservation of particles, this is also the fluctuation in the number of free electrons. Hence, if our goal is to detect a single added free electron due to a single photon, it is necessary that the noise fluctuation in the number of free carriers be less than unity. This means that $\delta N < 1$ and that N be even smaller

$$N < \frac{1}{[\log_2(f_2/f_1)]} \tag{7.33}$$

That is, the total number N of trapped electrons per time slot in the device must be less than unity. This is indeed a stringent demand on the purity of materials. The measure of its achievement is most likely to be found in the state of the art of MOS triodes since there is a high concentration of effort on the reduction of $1/f$ noise in these devices and since the technology of silicon is one of the most highly developed of all solid-state materials. The noise and photodetection properties of the MOS, operated with a floating gate, is completely parallel with that of the simple photoconductor which was discussed in the preceding section. The channel of the MOS plays the role of the photoconductor.

7.5. Photon Counting by MOS Triodes

Figure 7.9 shows schematically the parts of a floating-gate MOS (metal–oxide–semiconductor) triode. The channel electrons are induced by the positive bias charge on the gate. The voltage on the gate is a measure of the number of channel electrons and is

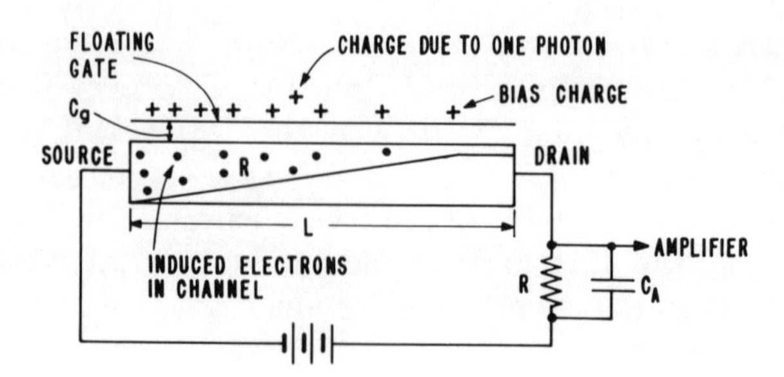

Fig. 7.9. Schematic diagram of a floating-gate MOS (metal–oxide–semiconductor) for detecting single charges or photons.

also equal to the potential drop along the channel. The channel current is accordingly proportional to the square of the voltage drop along the channel just as it would be for the space-charge-limited current in a photoconductor.

The problem is to determine the conditions under which the addition of a single charge on the gate can be detected. There are various ways that a single photogenerated charge can be added to the gate. In particular, the principle of charge transfer by charge-coupled electrodes can put a photogenerated charge on the gate for a fixed time τ and then remove it. During the time τ, the gate is floating. The effect of adding a single charge to the gate for a time τ is to add a single charge to the channel for the same time. But this operation is the complete equivalent of a photoexcitation in a photoconductor where a single extra electron is added to the conduction band for a time equal to the lifetime of a photocarrier. In both cases, MOS and photoconductor, the added charge causes a signal charge

$$Q_s = e\frac{\tau}{T_r} \tag{7.34}$$

to flow through the device. And in both cases, the signal charge must be detected against a background noise charge due to the resistance of the channel (or photoconductor)

$$Q_n = (2kTC_A)^{1/2} \tag{7.35}$$

where $RC_A = \tau$ is the input time constant of the amplifier to which the MOS or photoconductor is connected. With Eqs. (7.34) and (7.35) as the starting point, the conditions for the detection of a single photon are the same for the MOS as for the photoconductor, namely,

$$G \geqq \frac{1}{e}(2kTC_A)^{1/2} \tag{7.36}$$

Since*

$$G = \frac{\tau}{T_r} = \frac{RC_A}{RC_g} = \frac{C_A}{C_g} \tag{7.37}$$

where C_g is the gate capacitance, Eq. (7.36) can also be written

$$G \geqq \frac{2kTC_g}{e^2} \tag{7.38}$$

For $C_g = 10^{-14}$ farad, the gain G is $\geqq 4 \times 10^3$.

We conclude here again, as for the photoconductor, that if the only noise to be surmounted is the thermal noise of the channel resistance, the conditions for counting individual photons are easily met.

But the MOS also displays a $1/f$ noise spectrum and its effect, as for the photoconductor, is to mask the detection of single photons when the number of trapped electrons per "time slot" exceeds unity. There is a reasonably well documented source for $1/f$ noise in the MOS such that the number of trapped electrons per time slot can be given a simple meaning. The noise model was first proposed by McWhorter[(M-1)] and is shown schematically in Fig. 7.10 which shows a cross section of the channel and its adjacent SiO_2 layer. Also shown are electrons trapped in the volume of the oxide and at the energy level of the silicon conduction band. The trapped electrons communicate with the channel by tunneling. It turns out numerically that the tunneling probability (for a potential barrier of several volts) decreases by a factor of 2 for each increase in distance from the channel of about 2 Å, that is, for each atomic layer. The time required for tunneling is the lifetime appropriate to the traps on succeeding atomic layers, and corresponds to the series of τ's in Fig. 7.7. Hence, a uniform distribution of traps extending some 40 Å into the oxide can account

* Note that $T_r = L^2/V_g\mu$ and $RC_g = (L^2/V_gC_g\mu)\,C_g = L^2/V_g\mu$.

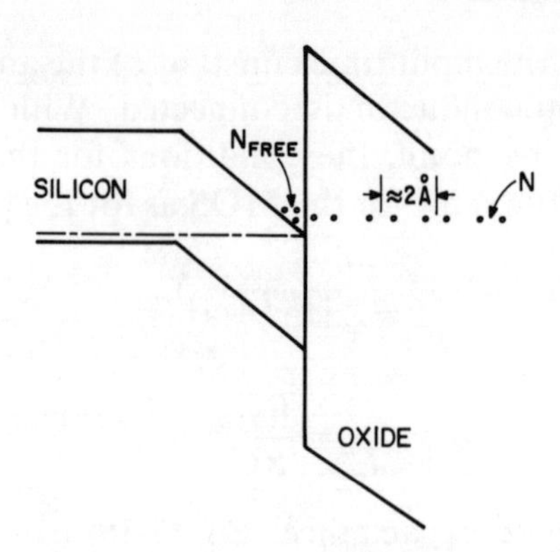

Fig. 7.10. Tunneling model for $1/f$ noise in an MOS structure.

for a $1/f$ noise spectrum extending over a frequency range of 2^{20}, or about 10^6.

Of particular interest is that the parameter N in Eq. (7.31) for the $1/f$ noise spectrum can be read directly as the total number of traps per atomic layer of oxide per energy slice kT for the area of channel facing the oxide. The criterion for photon counting, $N < 1$, then means a volume density of traps near the surface of the oxide less than $10^{14}/\text{cm}^3 \cdot kT$, or a surface density of traps per atomic layer of less than $10^6/\text{cm}^2 \cdot kT$. We have used a channel surface area in contact with the oxide of 10^{-6} cm^2. Some recent data of Fu and Sah[F-1] give measured trap densities near the oxide surface of 10^{14}–$10^{16}/\text{cm}^3 \cdot kT$ depending on the gate bias and, in some cases, on the distance from the oxide surface. Thus, the prospect of detecting single photons is at present marginal but not to be ruled out as the technology of silicon–silicon oxide interfaces improves.

A simpler and more accessible way of assessing the parameter N is to measure the frequency at which the $1/f$ noise spectrum intersects the thermal noise. This intersection is obtained by equating the thermal noise to the $1/f$ spectrum given by Eq. (7.31)

$$\frac{4kT}{R} = \frac{4V^2\mu^2e^2}{L^4f}N \tag{7.39}$$

We make use of the relations

$$R = \frac{L^2}{N_f e\mu}$$

and

$$N_f = \frac{VC_g}{e}$$

where N_f is the total number of electrons in the channel, to convert Eq. (7.39) into

$$f = \frac{Ve^2\mu}{kTL^2C_g} N \tag{7.40}$$

Note that Eq. (7.40) with $N = 1$ is also the criterion for detecting a single photon above the thermal noise [Eq. (7.38)] since a single trapped electron with a lifetime $\tau = 1/2f$ has the same effect as a single electronic charge held on the gate for a time τ. To confirm this we rewrite Eq. (7.40) as

$$\frac{V\mu}{2L^2f} = \frac{\tau}{T_r} = G = \frac{kTC_g}{2e^2} \tag{7.41}$$

Except for a numerical factor, which arose in part from the summing of the $1/f$ noise sources and does not appear in "the detection" of single photons, Eq. (7.41) matches Eq. (7.38).

It is useful to compute f in Eq. (7.40) for $N = 1$ and for a representative set of values for the remaining parameters: $V = 5$ volts, $\mu = 10^3$ cm^2/volt · sec, $L = 2 \times 10^{-3}$ cm, and $C_g = 10^{-13}$ farad. Then, $f = 10^5$/sec.

Data reported by Klaassens(K-1) on MOS devices of comparable dimensions show the $1/f$ spectrum intersecting the flat portion of the noise spectrum at about 10^5 Hz. Hence, the evidence again points to the present state of the art as being tangential to the refinement needed for the detection of single photons.(C-1)

7.6. Non-Photon-Counters

The photoconductor, shown in Fig. 7.4, and the MOS triodes are both capable, in principle, of detecting single photons. The phototransistor (Fig. 7.11) is fundamentally incapable of single photodetection.[R-2] The same would be true of a photoconductor if the states out of which the photoelectron was excited were located at the Fermi level instead of well below it, as shown in Fig. 7.4.

The reasons for these statements are the same reasons which govern the visibility of $1/f$ noise over and above thermal noise. The signal charge due to an extra electron, whether it be photo-excited or generated from one of the $1/f$ noise sources, increases linearly with the field across the device. In particular this charge is given by

$$Q_s = e\frac{\tau}{T_r} = e\frac{\tau V\mu}{L^2}$$

As the voltage is increased, the signal charge increases, but the thermal noise charge [Eq. (7.5)]

$$Q_n = (2kTC)^{1/2}$$

remains constant. At some voltage, then [see Eq. (7.39)], the signal charge exceeds the thermal-noise charge and single photons can be detected. The essence of this argument depends on the *extra signal electron being generated from a source which is different*

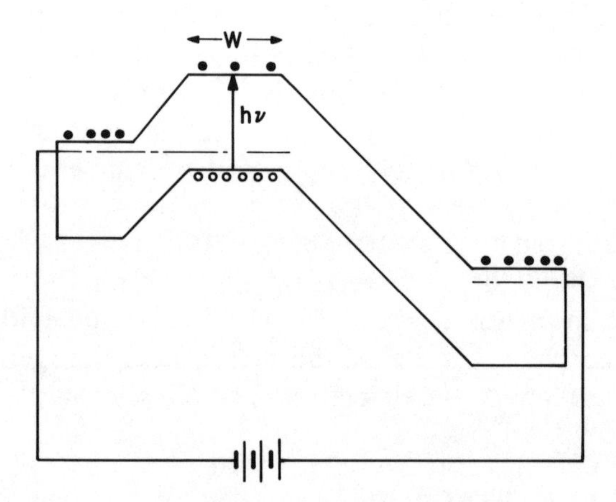

Fig. 7.11. Bipolar phototransistor showing a common origin for thermal electrons and photo-electrons in the base.

from and has a longer lifetime than the source for the thermal electrons. If, on the contrary, the signal electron comes from the same source as the thermal electrons, the applied field enhances the contribution of both to the same degree, since they both have the same lifetimes.

In the phototransistor, the thermal electrons in the base have the same lifetime as the photoexcited electron since they are both excited from the valence band. Hence, the noise charge as well as the signal charge is multiplied by the same gain factor

$$G = \frac{\tau}{T_r} \approx \frac{\tau k T \mu}{W^2 e}$$

and a signal electron can not be detected. In fact, it must compete with $N^{1/2}$ noise electrons where N is the number of electrons (minority carriers) in the base of the *n–p–n* transistor.

So also, in a photoconductor, if the photoelectron is excited out of states lying at the Fermi level, it has the same lifetime against recapture as the thermal electrons. Under an applied field, both the signal electron and the noise electrons are multiplied by the same gain factor

$$G = \frac{\tau}{T_r} = \frac{\tau V \mu}{L^2}$$

The number of noise electrons is $N^{1/2}$, where N is either the total number of free electrons or the total number of electrons in the states at the Fermi level, whichever is smaller.

In the photoconductor shown in Fig. 7.4, the thermal electrons originate from shallow donors and show negligible fluctuation in their numbers. Their effective lifetime, for noise purposes, is (see Section 7.2) their collision time with the lattice ($\sim 10^{-13}$ sec) which is much smaller than the lifetime of photoexcited electrons. The effective lifetime of the thermal electrons is their average time of residence in the category of electrons moving to the right or to the left. It is fluctuations in the numbers in these two categories that generate the thermal noise. Neither the lifetime nor the charge contribution el/L, where l is a mean free path, are affected by the applied field until the applied field is large enough to significantly enhance the thermal energy of the electrons.

7.7. Summary

Properly designed photoconductors and MOS triodes with floating gates can, in principle, detect single photons or single electron charges. In order to realize this counting ability, the sources of $1/f$ noise must be reduced to the level at which less than one "trapped" electron, whose lifetime lies in the neighborhood of the counting frequency, can be present in the device.

A model for $1/f$ noise is outlined in terms of particle concepts that are convenient for evaluating the effect of $1/f$ noise on the ability to detect single photons. The measure of the magnitude of $1/f$ noise is a parameter N defined as the number of "trapped" electrons per time slot. A time slot refers to a factor of 2 range in the lifetime of trapped electrons.

It is shown that devices in which the photoelectron is excited from the same states out of which the thermal electrons originate cannot, in principle, detect single photons unless the thermal noise charge $(2kTC)^{1/2}$ is less than the charge of one electron. At room temperature this leads to the unrealistic device capacitance of 10^{-18} farad or 10^{-6} cm.

7.8. References

C-1. J. E. Carnes and W. F. Kosonocky, Noise sources in charge-coupled devices, *RCA Rev.* **33**, 327–343 (1972).

F-1. H. S. Fu and C. T. Sah, Theory and experiments on surface $1/f$ noise, *IEEE Trans. Electron Devices* **ED-19**, 273–285 (1972).

K-1. F. M. Klaasen, Characterization of low $1/f$ noise in MOS transistors, *IEEE Trans. Electron Devices* **ED-18**, 887–891 (1971).

M-1. A.L. McWhorter, "$1/f$ Noise and Germanium Surface Properties," in *Semiconductor Surface Physics*, p. 207 (1956), Univ. of Pennsylvania Press, Philadelphia.

R-1. A. Rose, *Concepts in Photoconductivity and Allied Problems*, Chapt. 6 (1963), John Wiley & Sons, Inc., New York.

R-2. A. Rose, "An Analysis of Photoconductive Photon Counting," in *Proceedings of the Third International Photoconductivity Conference*, Stanford Univ., August, 1969, pp. 7–11 (1971), Pergamon Press, Oxford.

CHAPTER 8

SOLID-STATE PHOTOGRAPHIC SYSTEMS, LIGHT AMPLIFIERS, AND DISPLAY SYSTEMS

8.1. Introduction

There is clearly a redundancy in the title of this chapter. A photographic system, for example, is also a light amplifier and a display system. On the other hand, the literature tends to classify a wide range of devices according to whether their primary functions are photographic, amplifying, or display. The common trend in all of the devices is that, with few exceptions, the input light is absorbed by a layer of photoconductive material and the resultant changes in conductivity, charge, or voltage are used in conjunction with a second mechanism to form an image. This chapter explores the fundamental limitations on the sensitivity of such photoconductive layers.

Photoconductors have appeared to offer generous possibilities for high-sensitivity devices by virtue of the known fact that a weak incident stream of photons can give rise to a large stream of electrons passing through the photoconductor. This photoconductive gain (the ratio of the current of electrons to the current of photons) has been observed to reach values as high as 10^6 in some sensitive photoconductors such as cadmium sulfide. The attempts to utilize the high photoconductive gains in actual image-forming devices in order to reduce the required optical exposures have met with a stubborn lack of success. The reasons are fundamental and form the major content of this chapter.

8.2. Blocking Contacts

We begin with the simplest photoconductive arrangement, a photoconductor with blocking contacts (Fig. 8.1). Here a voltage is applied to the photoconductor via contacts that do not allow electrons or holes to enter the photoconductor. The same electrodes can, on the other hand, readily accept electrons and holes *from* the photoconductor. The result is that the absorption of one photon can, at most,* give rise to one electronic charge in the outside circuit, that is, a photoconductive gain of unity. In spite of the restriction to unity gain, this class of photoconductors has been widely and successfully used in imaging devices.

The blocking contacts can be formed in a variety of ways. The back-biased *p–n* junction is, perhaps, the best known structure (Fig. 8.1a). The use of metal contacts (Fig. 8.1b) with a high work function (potential barrier) at the cathode and a low work function at the anode can achieve the same results. The supply of electrons from the cathode and holes from the anode can be vanishingly small. An electrolytic contact (Fig. 8.1c), in the absence of chemical reactions at the electrode surfaces, forms one of the most effective blocking contacts. As is evident in the figure, the electrons are attached to the negative ions and lie energetically far below the conduction band. Similarly, the holes, which are attached to the positive ions, lie energetically far above the valence band. A variation on the principle of electrolytic contact is the use of a glow discharge to deposit negative ions, or positive ions, on the surface of the photoconductor (Fig. 8.1d). The highly blocking contacts achieved on zinc oxide in the Electrofax and on amorphous

* We postpone until the next chapter a discussion of the possibility that the photoexcited electron can, by impact ionization under high fields, give rise to a large number of secondary electrons.

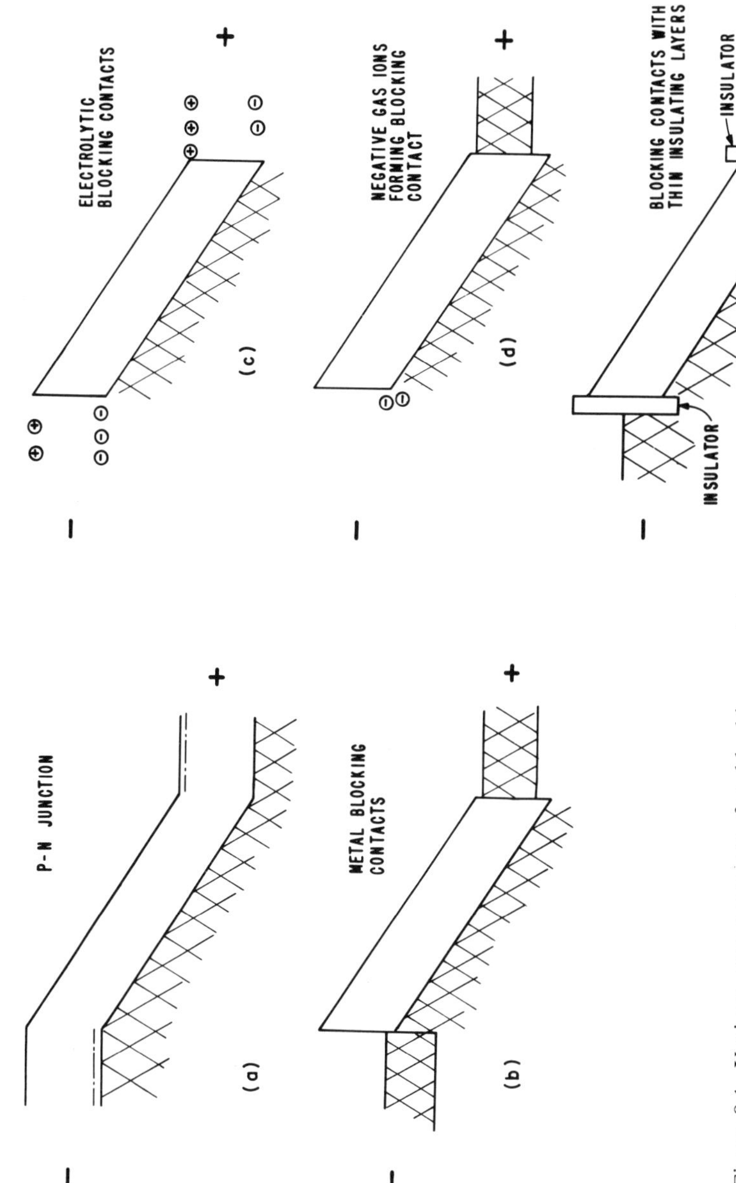

Figure 8.1. Various arrangements for blocking contacts. (a) Reverse-biased *p–n* junction, (b) metal–semiconductor contacts using high- and low-work-function metals, (c) electrolytic contacts, (d) gas-ion contact, (e) thin film of insulator.

selenium in the Xerographic arrangements for office copiers are classic examples. Finally, thin layers of insulator can obviously be used to form blocking contacts (Fig. 8.1c). Here, however, only transient, and not steady-state, photocurrents can be observed.

The virtue of the blocking contacts, as contrasted with the ohmic contacts discussed below, is that they can combine an arbitrarily high insulation (low dark current) with a fast response to light. The latter is determined by the transit time of charge carriers between the electrodes.

8.3. Sensitivity Aspects of Blocking Contacts

The photosensitivity of photoconductors with blocking contacts is limited to at most one electronic charge per absorbed photon. In a typical arrangement, shown in Fig. 8.2, a photoconductor and an adjacent insulating electrooptic material form a sandwich between two electrodes. In the absence of signal light on the photoconductor, the field is approximately the same in the photoconductor and electrooptic layer and is adjusted so that it is below the threshold for showing a significant electrooptic effect. The action of light on the photoconductor is to reduce the field in the photoconductor and increase it in the electrooptic layer.

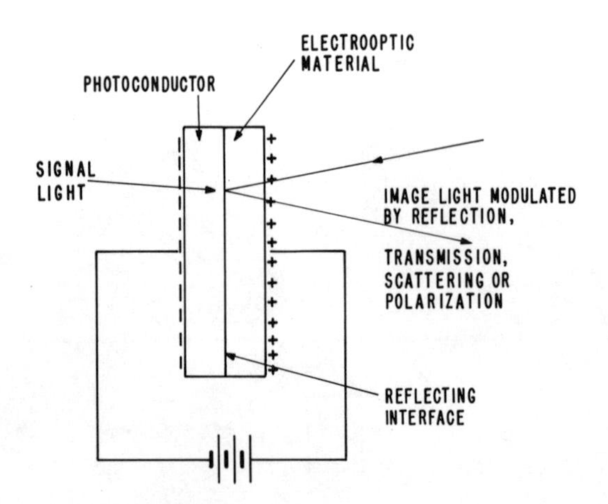

Fig. 8.2. Schematic arrangement for display device using a photoconductivity-controlled electrooptic material.

If the thickness and dielectric constants of the two layers are comparable, the number of photons required to transfer substantially all of the field to the electrooptic layer is given closely by the initial applied field itself. That is, the initial charge in electrons/cm^2 is

$$N = 6 \times 10^5 K\mathscr{E} \text{ electrons/cm}^2 \qquad (8.1)$$

where K is the dielectric constant and $\mathscr{E}$ the electric field in volts/cm. The photographic exposure in photons/cm^2 required to shift the field is also given by Eq. (8.1). Since electrooptic materials commonly use fields in the neighborhood of 10^6 volts/cm, the exposures are commonly of the order of 10^{13} photons/cm^2. This exposure is some 10^3 times larger than that needed for a sensitive photographic film. The photographic exposure itself is some 10^2 times larger than it would be if the effective quantum efficiency of film were 100 % rather than its current value of 1 %. This excess exposure does have certain virtues. It leads, in principle and in practice also, to very high resolution displays of the kind needed for holographic purposes.*

The direction in which higher sensitivity is to be achieved is clearly in the direction of lower electric fields. Either the total field across the sandwich must be reduced or the fraction of field needed to be transferred from photoconductor to electrooptic layer must be reduced. In either case, the electrooptic layer must be capable of operating at significantly lower fields or changes in field.

The office copying arrangements (Fig. 8.3) using amorphous selenium (Xerox)[S-2] or fine-grained zinc oxide powders (Electrofax)[Y-1] both operate with initial fields of about 10^5 volts/cm and

* Note that in this chapter we are using the term sensitivity in its common meaning, namely, the reciprocal of the magnitude of the photographic exposure. It is *not* the same as quantum efficiency. In order to determine the quantum efficiency, we would need to know the signal-to-noise ratios of the various display devices. These data are not yet available. A low sensitivity (large exposure) in this chapter may very well correspond to a high quantum efficiency if the resultant display has a high signal-to-noise ratio and correspondingly high resolution. See, for example, T. L. Credelle and F. W. Spong, Thermoplastic media for holographic recording, *RCA Rev.* **33**, 206–225 (1972) for an illustration of the combination of high resolution, high quantum efficiency, and large photographic exposures.

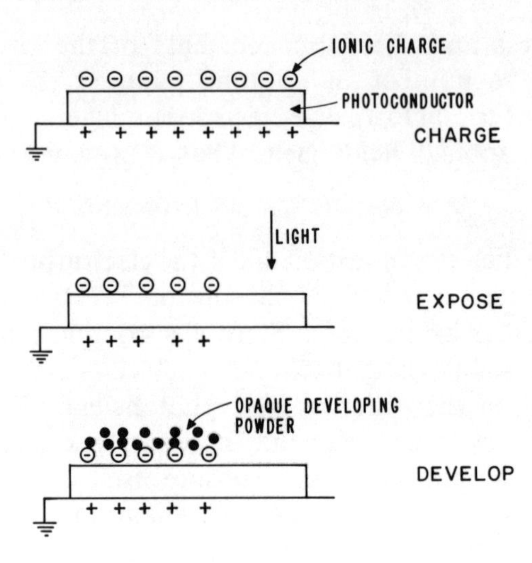

Fig. 8.3. Electrophotographic system for office copiers.

with exposures that reduce these fields to substantially zero. Their exposures are normally about 10^{12} photons/cm^2. Some improvement in sensitivity has been achieved by discharging only a fraction of the initial charge and developing the resultant charge pattern using a null method[D-1] (Fig. 8.4).

Liquid-crystal light values (Fig. 8.5),[H-1,S-2] operate with fields in the range of about 10^4 volts/cm and should operate with exposures of about 10^{11} photons/cm^2, that is, well below the range of office copiers.

The most sensitive use of photoconductors with blocking contacts is to be found in the television camera tubes of the vidicon class (see Chapter 3). Here the field across the photoconductor ranges from 10^4 to 10^5 volts/cm. However, the change in field required to transmit a good picture is only 10^3 volts/cm, or about 1 volt across a layer 10 μm thick. While the exposure, 10^{10} photons/cm^2, is comparable with that for sensitive photographic film, the signal-to-noise ratio of the transmitted picture is several times larger than that for the film and corresponds to an effective quantum efficiency of about 10%. The photo process, itself, has a quantum efficiency of close to 100%. The noise, however, is that of the

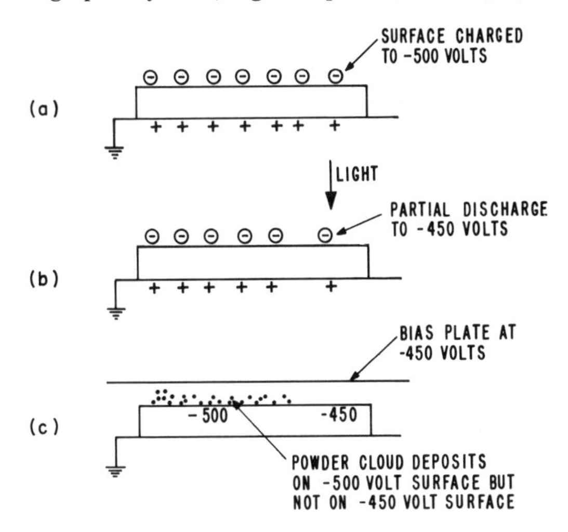

Fig. 8.4. A null-system variation of Fig. 8.3.

amplifier and is several times larger than the noise inherent in the photon flux.

The photoconductors that have shown good performance in the blocking contact mode in vidicons* are lead oxide, amorphous selenium, back-biased silicon *p–n* junction arrays, and, more recently, cadmium selenide with a thin layer of insulator which blocks the entrance of electrons from the beam but transmits

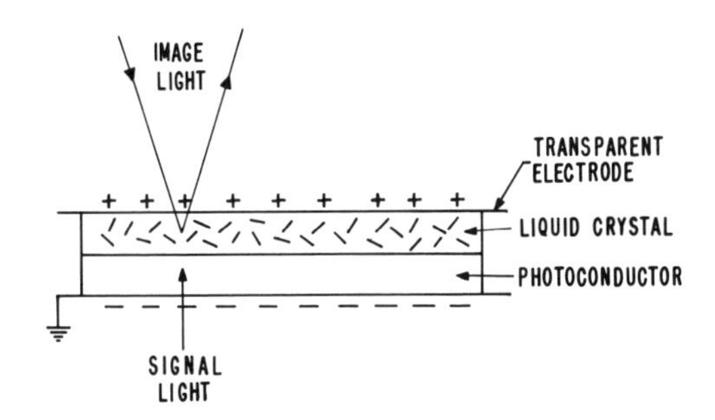

Fig. 8.5. Display system using liquid-crystal light scatterer.

* See Chapter 3 for references.

holes leaving the cadmium-selenide layer. The most widely used material, currently, is lead oxide. In the case of lead oxide and amorphous selenium, the surface layers exposed to the scanning beam must contain deep lying electron traps which capture the beam electrons and prevent them from entering the volume of the layer.

8.4. Ohmic Contacts

The use of ohmic contacts on a photoconductor permits the achievement of extremely high current gains or charge gains. As many as 10^6 electrons can be passed through the photoconductor per absorbed photon. Figure 8.6 shows the nature of an ohmic contact and the principle of the photoconductive gain.

In contrast to the blocking contact, the ohmic contact supplies a reservoir of electrons ready to enter the photoconductor as needed. In the dark, the current of electrons leaving the photoconductor is equal to the number of electrons in the photoconductor divided by their transit time. The ohmic contact replaces these electrons at the same rate at which they leave so that the density in the photoconductor is maintained constant. If the voltage across the photoconductor is, for example, doubled, the electrons leave at twice the previous rate, and the ohmic contact supplies the replacements at twice the previous rate. At this point one might ask what keeps the ohmic contact from supplying more

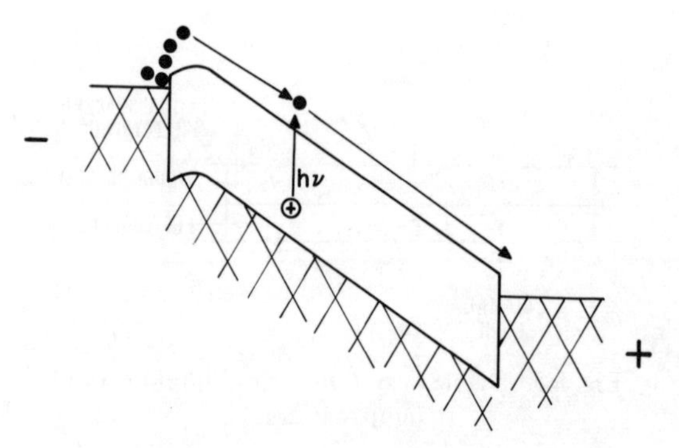

Fig. 8.6. Model for photoconductor with photoconductive gains greater than unity.

electrons than are needed to support the ohmic currents. What keeps the reservoir of electrons at the cathode from entering the volume of the photoconductor and making it highly conducting?

The answer is that an excess of electrons does, indeed, begin to enter the photoconductor. To the extent that these extra electrons enter, they represent a negative charge in the previously neutral photoconductor. A charge in the volume of any material is dissipated in an ohmic relaxation time given by the conductivity of the material.

$$\tau_{\text{rel}} \doteq 10^{-12} \frac{K}{4\pi\sigma} \text{ sec} \tag{8.2}$$

where σ in the conductivity in $(\text{ohm} \cdot \text{cm})^{-1}$ and K is the dielectric constant. If the material is not connected to any fixed potential source, the volume charge ends up at the surface of the material in the relaxation time and is so distributed as to yield zero electric field within the volume. If the material is contacted by one or more electrodes connected to ground or to a battery, the charge exits via these electrodes. Thus, as the excess charge tries to enter the volume from the reservoir at the cathode, it is dissipated to the electrodes in a relaxation time. If the relaxation time is small compared with the transit time, the charge is rapidly dissipated before it makes any significant penetration of the interelectrode spacing. In this way, the ohmic contact is able to supply exactly as many electrons as are needed to replace those leaving the photoconductor due to an ohmic flow of charge.

The pattern outlined above describes the so-called ohmic range of currents—a range in which the current is proportional to the voltage.

If, now, the voltage is increased to the point that the transit time becomes less than the relaxation time, an excess charge from the reservoir does, indeed, enter and occupy the volume of the photoconductor. The total amount of this charge Ne is governed by the field at the anode by a relation like that of Eq. (8.1)

$$Ne = \frac{K}{4\pi} \mathscr{E}$$

$$N = 6 \times 10^5 K\mathscr{E} \text{ electrons/cm}^2 \tag{8.3}$$

As in Eq. (8.1), this is the sheet charge density required to terminate the field lines from the anode. In this regime, where the reservoir supplies excess charge to the interior, the current increases as the square of the voltage since the voltage both decreases the transit time and increases the volume density of charge carriers. The currents in this regime are called space-charge-limited currents.

The phenomenon of space-charge-limited currents imposes an important constraint on the operation of photoconductors.[R-1] Usually a system is designed so that, in the dark, the photoconductor has sufficient insulation to store charges for a significant time, often in the range of seconds. The effect of light is then to release the charge in the lighted areas while the charge is still maintained in the dark areas. For this reason, a photoconductor whose resistivity has been chosen properly for charge storage in the dark can not be operated at fields high enough to invoke space-charge-limited current flow. At such high fields, the conductivity of the photoconductor is increased by the applied field to the point that the charge pattern leaks off in the dark in times too short for the proper functioning of the device. This constraint on the maximum applied field becomes also a constraint on the maximum photoconductive gain.

Returning to Fig. 8.6, the mechanism of photoconductive gain is shown schematically. A free electron is created by a photon from some fixed center. The free electron passes to the anode leaving behind a positive charge. The positive charge is neutralized by an extra electron from the reservoir at the cathode. The replacement electron also passes to the anode and is followed by another replacement electron. This process continues until one of the free electrons is captured by the fixed positive center.*

The number of charges passed per photon is the photoconductive gain and is

$$G = \frac{\tau}{T} \tag{8.4}$$

* A more accurate description leading to the same result is that a statistical excess of one free electron remains in the neighborhood of its point of generation for a time equal to its lifetime. During this period τ/T, extra charges are passed through the photoconductor owing to its increased conductivity.

where τ is the lifetime of the extra electron or, what is equivalent, the lifetime of the positive center. T is the transit time of a free electron and

$$T = \frac{L}{\mathscr{E}\mu} \tag{8.5}$$

From Eq. (8.3), the maximum field that can be applied short of the space-charge-current regime is

$$\mathscr{E} = \frac{4\pi}{K} Ne = \frac{4\pi}{K} neL \tag{8.6}$$

where n is the volume density of free electrons in the dark. Combination of Eqs. (8.4), (8.5), and (8.6) then yields for the maximum gain, short of the space-charge-current regime

$$G = \frac{\tau}{T} = \frac{\tau}{(4\pi/K)\, ne\mu} = \frac{\tau}{(4\pi/K)\, \sigma} = \frac{\tau}{\tau_{\text{rel}}} \tag{8.7}$$

In Eq. (8.7), τ is both the lifetime of a photoexcited electron and the response time of the photoconductor to light. That is, in the absence of shallow traps, the photocurrent will reach a steady-state value in τ sec after the start of illumination.

The introduction of shallow traps has two effects, as already discussed in Chapter 7 [see Eq. (7.24)]. The shallow traps increase the response time relative to the lifetime so that

$$\tau = \tau_{\text{resp}} \frac{n}{n_t} \tag{8.8}$$

where n_t is the density of shallow-trapped electrons. Also, the shallow traps allow a higher applied field before the advent of space-charge-limited currents. The applied field must now be sufficient to double not only the density of free electrons, but also the density of trapped electrons which are in thermal equilibrium with the free electrons. Hence, at this higher field,

$$T = \tau_{\text{rel}} \frac{n}{n_t} \tag{8.9}$$

Insertion of Eqs. (8.8) and (8.9) into Eq. (8.4) then yields

$$G = \frac{\tau_{\text{resp}}}{\tau_{\text{rel}}} \tag{8.10}$$

That is, the maximum gain before the advent of space-charge-limited currents is the ratio of the response time of the photoconductor to its ohmic relaxation time, just as in Eq. (8.7) for a trapfree material. There are special trap distributions that can yield a somewhat larger gain than that given by Eq. (8.10). We will return to these later. For the present, Eq. (8.10) forms the basis for discussing the sensitivity of imaging devices using photoconductors with ohmic contacts.

8.5. Sensitivity Aspects of Ohmic Contacts

We will consider several representative arrangements using photoconductive layers having ohmic contacts and show that their sensitivity under the constraints of actual use is the same as that achieved by photoconductive layers having blocking contacts.[R-2] Sensitivity is measured here by the optical exposure in photons/cm^2 required to form a picture. This will turn out to be given by the electric field (or a fraction of it) across the photoconductor, just as was the case for blocking contacts.

The simplest example is the photoconductive layer used in the vidicon type of television camera tube. This layer must satisfy two conditions. The first condition is that its insulation in the dark must be sufficient to store charges for the time required to scan the target with the scanning beam. In this way, the full complement of charges accumulated by the light on any picture element in the time between successive scans is retained. The second condition to be satisfied is that the photoconductivity excited by the light must not persist more than one scan time (1/30 sec). If it did persist for longer times, the camera would not be performing its functions of providing 30 independent pictures per second. Objects in motion would be objectionably blurred.

These two conditions define the parameters in Eq. (8.10). The first condition requires that the dielectric relaxation time τ_{rel} be such that

$$\tau_{rel} \geqq \tfrac{1}{30} \text{ sec}$$

The second condition requires that the response time of the photoconductor be such that

$$\tau_{resp} \leqq \tfrac{1}{30} \text{ sec}$$

The combination of these two conditions then yield the photoconductive gain

$$G \leqq 1$$

Stated in other terms, the use of a photoconductor with a gain greater than unity would lead to response times longer than 1/30 sec and would transmit objectionably blurred pictures.

These conclusions have been amply confirmed by numerous unsuccessful attempts to use high-gain photoconductors in vidicons. The only successful vidicon making use of a photoconductor with ohmic contacts is that in which Sb_2S_3 forms the photosensitive target. Here, the operating sensitivity corresponds at best to a photoconductive gain of unity. The optical exposure for a reasonably good picture is in the order of 10^{10} photons/cm^2 corresponding to a change in field of 10^3 volts/cm.

A second class of devices makes use of a photoconductive resistance in series with some picture-producing layer or process. We use the term resistance in contrast to a charge-storage layer. The resistivity is too low for the full storage of charge during the time of the optical exposure. What is stored for this period is the photoconductive state, namely, the enhanced conductivity provoked by light.

One example of this arrangement is the light amplifier consisting of a photoconductive layer in series with an electroluminescent layer.[(K-1)] An ac voltage is applied across the combination. In the dark, the ac voltage is predominantly across the photoconductor. Under illumination, it is shifted to the electroluminescent layer. The impedance of the latter is largely capacitive. Another example is the combination of a photoconductive layer in series with a liquid-crystal light valve. The liquid-crystal light valve in this case is thought to vary its reflectivity or transmission as a consequence of a turbulence provoked by the passage of current. A third example is the photographic process in which the photoconductive layer is in series with an electroplating solution.* The incidence of light on the photoconductor permits an electro-

* A commercial arrangement in which silver is plated on to the surface of a layer of zinc-oxide photoconductor has been developed by Minnesota Mining and Manufacturing Co.

plating current to deposit metal ions on the surface of the photoconductor.

In each of these examples, it is clear that, in order to get a picture of reasonable contrast, the optical exposure must at least double the dark current in the photoconductor. In a linear device, one for which the light output is proportional to the current through the device, the doubling of the current by light would yield a contrast of light to dark areas of only 2:1. For a nonlinear device in which the light output varies as I^n, a doubling of the current would yield a contrast of n:1. Since the contrast in a good quality picture is likely to be in excess of 10:1, it is unlikely that this can be achieved with exposures less than that needed to double the dark current. To do so would require a nonlinearity such that the light output varied as I^n, where $n \gg 10$. Furthermore, the onset of the nonlinearity would have to be uniform over the surface of the device to the extent that the onset current could not deviate by more than the fraction $1/n$. While all of this is, in principle, possible, none of the manifold arrangements for making pictures has approached such high values of n and such highly uniform values of nonlinearities.

Accordingly, our estimate for the optical exposure required to form a picture is based on a doubling of the dark current.

We begin with the assumption that the electric field across the photoconductor is just sufficient for the onset of space-charge-limited currents. In this way one can achieve the highest photoconductive gain without reducing the resistivity of the photoconductor. The electric field [see discussion of Eqs. (8.8)–(8.10)] is then related to the total free and trapped charge by

$$N = (n + n_t)\,L = 6 \times 10^5 K\mathscr{E} \text{ electrons/cm}^2 \qquad (8.11)$$

where n and n_t are the volume densities of free and trapped electrons, respectively, and N is their equivalent sheet density for a photoconductor of thickness L.

But, Eq. (8.11) also defines the exposure in photons/cm^2 required to double the dark current since the number of photons must be sufficient to double both the free- and trapped-electron densities. Hence, as in the case of blocking contacts, the optical exposure is determined by the electric field across the photoconductor. The optical exposure does not depend on and is not reduced by the high gain properties of the photoconductor.

If the field across the photoconductor were less than that required for the onset of space-charge-limited currents, the optical exposure required to double the dark current would not be altered, but the photoconductive gain would, of course, be reduced. Fields higher than this threshold are ruled out since, by assumption, the resistivity of the photoconductor was chosen to fit the constraints of the device and can not be altered.

We turn now to a third category of devices in which a relatively insulating photoconductor and an insulating electrooptic material form a sandwich between two electrodes. The electrooptic material can be a field-dependent doubly-refracting material, a ferroelectric, or a deformable organic layer or other light-modulating material. The major features are that the field in the dark is below threshold for the electrooptic material. Under illumination, the field across the photoconductor is relaxed and that across the electrooptic material is increased. Here, again, the argument for an optical exposure which at least doubles the dark conductivity of the photoconductor is cogent. Let the time for the optical exposure be t. During this time it is essential that the field across the photoconductor not relax in the dark areas of the picture. If the dark relaxation time is much greater than t, the light must correspondingly increase the dark conductivity by much more than a factor of 2. The optimum condition exists when the dark relaxation is, perhaps, $2t$. The illuminated areas then relax in a time t or less. The argument for the minimal optical exposure then follows that already outlined for the current-sensitive devices. The exposure is again given by Eq. (8.11) and is the same as that for a photoconductor with blocking contacts.

It should be noted that, in the present arrangement of a photoconductor adjacent to a layer of insulating material, there is an additional constraint on the maximum achievable photoconductive gain. We recall that in the normal process of photoconductive gains, the extra, or photoexcited, electrons that are passed through the photoconductor are absorbed at the anode. Hence, the effect of the light is to increase the current through the photoconductor owing to an increase of conductivity at constant field. The increased current continues for the lifetime of the photocarriers and yields an extra charge. The ratio of the number of extra charges to the number of photons is the photoconductive gain ($G = \tau/T$).

If the extra electrons are not absorbed at the anode electrode, but remain at the anode surface of the photoconductor, they tend to reduce the field across the photoconductor and to nullify the gain process.

Consider the extreme case in Fig. 8.7a where the anode surface is floating and its capacitance to the anode electrode is vanishingly small. We introduce a pulse of light which increases the density of electrons by 10%. The photoconductor is assumed to be operating at the threshold for space-charge-limited currents where the transit time and dielectric relaxation time are equal in a trapfree material.

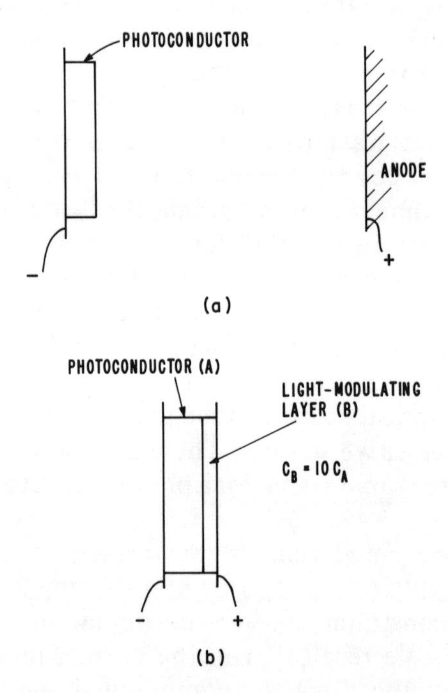

Fig. 8.7. Photoconductor with floating surface and (a) with low capacitance to anode, (b) with high capacitance to anode.

Hence, after one transit time, the field is reduced by an amount 10% more than would have occurred in the absence of light. This extra reduction was a result of the extra electrons lodged at the anode surface whose field lines terminate at the cathode. The result, then, is that after one transit time the extra current generated by the pulse of light has vanished because the 10% increase in conductivity was opposed by a 10% reduction in field. In this example the photoconductive gain is only 1.

In Fig. 8.7b the capacitance of the anode surface of the photoconductor to the anode electrode is 10 times larger than its capacitance to the cathode. Under these conditions, the 10% reduction in field, described for Fig. 8.7a, now requires 10 times as much charge. Hence, the extra current provoked by the light pulse will persist for 10 transit times and the gain process will terminate at a gain of 10.

In summary, the maximum achievable photoconductive gain is

$$G_{\max} = 1 + \frac{C_{\text{Light-modulating layer}}}{C_{\text{photoconductor}}} \tag{8.12}$$

More formal arguments for this conclusion have been given by H. S. Sommers[(S-4)] and P. J. Melz.[(M-1)] Note that the required exposure is still given by the applied field [Eq. (8.11)] even when the gain exceeds unity.

8.6. Special Arrangements

We have thus far shown that the optical exposure required for the usual photoconductors with ohmic contacts is the same as that for photoconductors with blocking contacts. The "usual" photoconductor is defined by Eq. (8.10) as having a maximum gain

$$G = \frac{\tau_{\text{resp}}}{\tau_{\text{rel}}} \tag{8.10}$$

Equation (8.10) was based on a material having shallow traps lying at and above the Fermi level and having a separate set of deep-lying states below the Fermi level out of which the photoelectrons are excited. If the states out of which the electrons are excited lie at the Fermi level and if their density exceeds the density of shallow traps, which are located in an energy range kT just above the Fermi level, by a factor M, then it has been shown[R-3] that Eq. (8.10) should read

$$G = M \frac{\tau_{\text{resp}}}{\tau_{\text{rel}}} \tag{8.13}$$

In brief, the performance is increased by the factor M and the optical exposures we have computed for ohmic contacts will be reduced by the factor $1/M$.

We will not reproduce the arguments leading to Eq. (8.13). The trap distributions needed for M values well in excess of unity are not likely to be found in other than highly purified materials such as germanium and silicon. Even for these materials, a more likely way of achieving M values greater than unity is via the use of the phototransistor arrangement shown in Fig. 8.8. The M value for this arrangement is in the order of $e(VV_0)^{1/2}/kT$ where the potentials V and V_0 are indicated in the figure.

The beam scanning process in the vidicon offers another special opportunity for increasing the sensitivity of these devices over and above the nominal photoconductive gain of unity. Normally, when these targets are scanned, the electron beam neutralizes the accumulated positive charge by depositing all of its electrons at the scanned surface. And normally the scanned surface is processed to insure this deposition. It is possible to process the scanned surface so that (see Fig. 8.9) for every electron deposited at the surface a number of electrons pass through the surface to the signal plate. This number then constitutes an extra gain factor and increases the sensitivity of the device.[S-4] The processing is particularly delicate to achieve high gains and to insure the high degree of uniformity over the surface required for good-quality pictures.

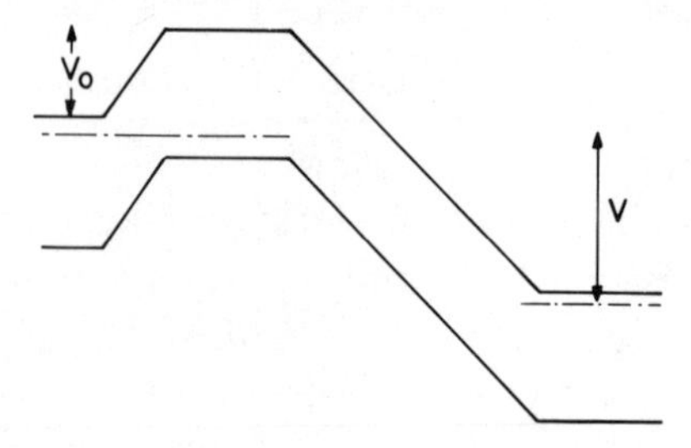

Fig. 8.8. Phototransistor.

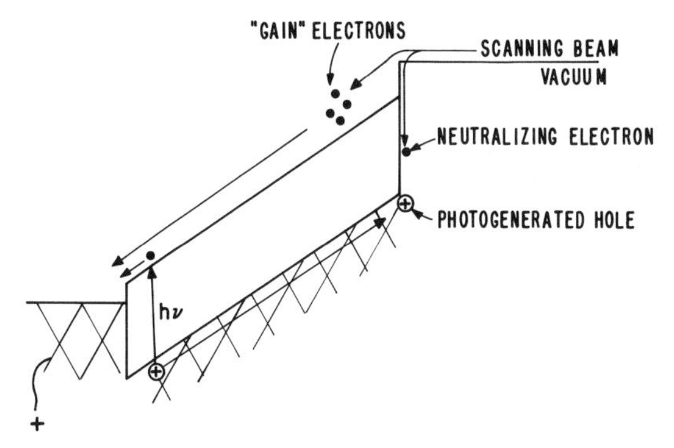

Fig. 8.9. Mechanism for obtaining gains greater than unity during the scanning of a vidicon target.

Evidence for modest gains achieved in this way was reported by Shimizu and Kiuchi.(S-3)

8.7. Null Systems

The major emphasis in this chapter has been on systems whose light-modulating properties depend on the total field or voltage across the photoconductor. These systems require optical exposures of the order of 6×10^5 $K\mathscr{E}$ photons/cm^2. The commonly encountered fields of 10^5 volts/cm then lead to exposures of 10^{12} photons/cm^2 which are some 10^4 times larger than that needed by an ideal photon-noise-limited device to present a "good" noisefree picture. The latter exposures would correspond to fields of only 10 volts/cm which are usually too low to ensure the transport of photocarriers across the photosensitive layer.

The combination of high fields for the transport of photocarriers and low fields for the sensitive generation of images can be achieved by any of a number of null systems. Passing mention was made of one such system in the form of an electrophotographic layer with an adjacent electrode held at the same potential as the surface of the photoconductor. Small changes in the surface potential of the photoconductor then controlled the deposition of a developing powder. Another null system is constituted by the low-velocity scanning beam of a vidicon. It responds to small changes in the large total voltage across the target.

The photogenerated charge pattern on the surface of a charged, photoconductive insulator contains all of the information in the incident photon flux. It is not unreasonable to expect that some null system will be developed for reading out this information efficiently at the small exposures required to achieve the ultimate in visual sensitivity.

Even now, charge patterns, down to individual electrons, can in principle be read out by scanning the surface with a probe whose probe tip consists of the floating gate of an MOS. This process is just the inverse of that already discussed in Chapter 7, where the charge pattern was moved past the MOS probe.

8.8. Summary

The optical exposure required for a photoconductively-controlled picture-forming device is normally the same for photoconductors with ohmic contacts and high photoconductive gains as for photoconductors with blocking contacts and unity gain. This exposure is usually $6 \times 10^5 K\mathscr{E}$ photons/cm^2, where $\mathscr{E}$ is the electric field across the photoconductors. Proportionately lower exposures are achieved when only a fraction of the field across the photoconductor needs to be transferred to the light-modulating layer. This can be achieved by a variety of null systems.

Some special trap distributions can, in principle, yield lower exposures for ohmically contacted photoconductors than for blocking contacts.

For the commonly uncountered fields of 10^5 volts/cm, the optical exposure is 10^{12} photons/cm^2 and is comparable with that needed for extremely fine-grained slow-speed photographic emulsions.

8.9. References

D-1. J. H. Dessauer and H. E. Clark, *Xerography and Related Processes* (1965), Focal Press, New York.

H-1. G. H. Heilmeier, L. A. Zanoni, and L. A. Barton, Dynamic scattering: A new electro-optic effect in certain classes of nematic liquid crystals, *Proc. IEEE* **56**, 1162–1171 (1968).

K-1. B. Kazan and F. H. Nicoll, An electroluminescent light amplifying panel, *Proc. IRE* **43**, 1888–1897 (1955).

M-1. P. J. Melz, Gain in electrophotography, *IEEE Trans. Electron Devices* **ED-19**, 433 (1972).

R-1. A. Rose, *Concepts in Photoconductivity and Allied Problems* (1963), John Wiley & Sons, Inc., New York.

R-2. A. Rose, The role of space-charge-limited currents in photoconductivity-controlled devices, *IEEE Trans. Electron Devices* **ED-19**, 430–433 (1972).

R-3. A. Rose and M. A. Lampert, Photoconductor performance, space-charge currents, and the steady-state Fermi level, *Phys. Rev.* **113**, 1227–1235 (1959).

S-1. M. Schadt and W. Helfrich, Voltage-dependent optical activity of a twisted nematic liquid crystal, *Appl. Phys. Letters* **18**, 127–128 (1971).

S-2. R. M. Schaffert and C. D. Oughten, Xerography: A new principle of photography and graphic reproduction, *J. Opt. Soc. Am.* **38**, 991–998 (1948).

S-3. K. Shimizu and Y. Kiuchi, Characteristics of the new vidicon-type camera tube using CdSe as a target material, *Japan. J. Appl. Phys.* **6**, 1089–1095 (1967).

S-4. H. S. Sommers, Jr., Response of photoconducting imaging devices with floating electrodes, *J. Appl. Phys.* **34**, 2923–2934 (1963).

Y-1. C. J. Young and H. G. Greig, Electrofax: Direct electrophotographic printing on paper, *RCA Rev.* **15**, 469–484 (1954).

General

J. H. Dessauer and H. E. Clark, *Xerography and Related Processes* (1965), Focal Press, New York.

M. A. Lampert and P. Mark, *Current Injection in Solids* (1970), Academic Press, New York.

A. Rose, *Concepts in Photoconductivity and Allied Problems* (1963), John Wiley & Sons, Inc., New York.

R. M. Schaffert, *Electrophotography* (1965), Focal Press, New York.

H. V. Soule, *Electro-Optical Photography at Low Illumination Levels,* Chapt. 9 (1968), John Wiley & Sons, Inc., New York.

CHAPTER 9

SOLID-STATE PHOTOMULTIPLIERS

9.1. Introduction

The vacuum photomultiplier has had a dramatic impact on the field of sensors owing to its ability to detect single photons. The vacuum photomultiplier, however, is a single-picture-element device. The same principle of multiplication has been extended to large-area image sensors by the several forms of image multipliers discussed in Chapter 6. These image multipliers are beginning to service a wide range of applications where the supply of photons is limited.

As powerful as these vacuum devices have proved to be in the field of sensing, they still leave open several significant areas for improvements. Their quantum efficiency is in the order of 10% rather than 100%. Their size, measured in inches, is larger than necessary. Their voltage sources range into the tens of kilovolts. And, finally, the extension of their spectral response into the micron wavelengths of the infrared spectrum is difficult. The last item, however, is beginning to yield to the hybrid combination of semiconductor photoemitters in vacuum (see Chapt. 6).

When we examine these areas in terms of solid-state photomultipliers, the formal possibilities for improvement are impressive. Solid-state sensors already operate with 100% quantum efficiency, not only in the visible but in the far-infrared range ($\sim 10\,\mu$m) as well. A solid-state photomultiplier must, almost of necessity, be in the range of micron thicknesses rather than inches. The voltage across these thicknesses can not exceed a few hundred to a thousand volts. In brief, a solid-state photomultiplier, if successful, would achieve the ultimate not only in sensitivity but, as well, in compactness and convenience.

Ironically, solid-state multiplication is one of the oldest recognized solid-state electronic phenomena. The early work of both Fröhlich[F-2] and von Hippel[H-1] in the 1930's argued strongly that dielectric breakdown was a consequence of the avalanche multiplication of single electrons starting from the cathode and giving rise to a surge of free carriers near the anode. In rough outline, they argued that an electron in a field of 10^6 volts/cm could gain a few volts of energy in a few hundred angstroms. It could then create a free electron and hole by impact ionization across the forbidden gap. If this process were repeated every few hundred angstroms, the single electron would give rise, in a micron of path length, to $\sim 10^{10}$ free carriers. In 2 μm of path length, this number would be 10^{20} and easily sufficient to cause dielectric breakdown by thermal destruction or decomposition of the insulating material. The impressive fact is that, once impact ionization occurs, it can lead to an enormous multiplication of carriers in distances of only a few microns. Moreover, if these theories of dielectric breakdown were correct, solid-state electron multiplication was occurring every day in the laboratory – whenever an engineer inadvertantly applied an overvoltage to a condenser and observed an electrical breakdown.

The only missing link between these primitive observations of breakdown and a highly sophisticated image sensor was the achievement of high values of multiplication *under control.* It would be necessary, for example, to be able to apply a high electric field across a thin insulating film of photoconducting material such that, in the dark, substantially no electrons would be present to initiate chains of multiplication. At this point, the introduction of a sprinkling of electrons at the cathode owing to the incidence of a sprinkling of photons would give rise to easily measurable pulses of charge at the anode. Single photons would be rendered easily observable. Indeed, the present system for office copiers using thin insulating layers of amorphous selenium (Xerography) or zinc oxide powder (Electrofax) should be ideal systems in which to make these observations. Not only are these materials excellent insulators, but also the method of charging by a corona discharge insures that the highest fields (owing to the blocking character of the ionic charge) could be achieved. *Thus far, however, no such multiplication has been observed.*[K-1]

There are several fundamental reasons that are responsible for the notable lack of success in achieving solid-state photomultipliers in insulating materials. First, the high electric fields required for impact ionization are also high enough to invoke the injection of electrons from the cathode by a tunneling mechanism. In fact, it is highly likely that the phenomenon of dielectric breakdown is initiated by these field-injected carriers rather than by avalanche multiplication.[W-5]

Second, the multiplication process in insulators is not easily controllable. Experimentally, the breakdown of insulators occurs at a sharply defined threshold field. Theoretically, the onset of impact ionization in polar insulators *should* be almost catastrophic.* This is in contrast to certain covalent semiconductors like silicon and germanium where there is some margin of control for the degree of multiplication.

The physical basis for this pessimistic appraisal of the prospects for solid-state photomultipliers in insulating materials will be discussed below. First, however, we will outline a somewhat less pessimistic appraisal of the state of the art of multiplication in semiconductors, and later discuss the reasons for the difference between insulators and semiconductors.

9.2. Multiplication in Semiconductors

The semiconductors in which electron multiplication has been observed under reasonable control are germanium, silicon, gallium

* At some critical field, the energy of the electrons jumps from a few tenths of a volt to the several volts (bandgap energy) required for impact ionization. The mean free path for impact ionization is only of the order of 10^{-7}–10^{-6} cm. Hence, in a layer a micron or more thick, these electrons will suffer over a hundred multiplications and lead to currents of catastrophic magnitude. By the same argument, layers less than a few thousand angstroms thick should tend to avoid this catastrophic behavior.

arsenide, gallium phosphide, and indium antimonide; and even in these few materials, reliable observations could be made only after the art of growing crystals free from dislocations could be achieved.

The usual measurements (Fig. 9.1) are made on back-biased p–n junctions.[(L-1)] Electrons are introduced at the cathode side of the junction or holes at the anode side by means of a light probe. At low fields, the current across the junction is constant, independent of the field, and equal to the photoexcited current of carriers. As the field is increased, the junction current (Fig. 9.2) increases owing to the impact ionization and consequent multiplication of carriers crossing the junction. At each field, a parameter α, the reciprocal of the average distance traveled by an electron or hole before making a free pair, is computed. A representative set of curves for α *versus* electric field for electrons and holes in silicon is shown in Fig. 9.3.

Two features are significant. In Fig. 9.3, the fields at which significant multiplication takes place are in the range of 10^5–10^6 volts/cm. These are relatively high fields in the sense that the competitive process of field emission or tunneling begins to be significant. More insulating materials, having larger forbidden gaps, tend to require even higher fields for significant multiplication and tend to be even more dominated by field-emission processes.

A second feature is the increasing steepness of the current–voltage curve (Fig. 9.2) as the multiplication exceeds a factor of 5 or 10. The steepness of this curve makes it almost impossible to

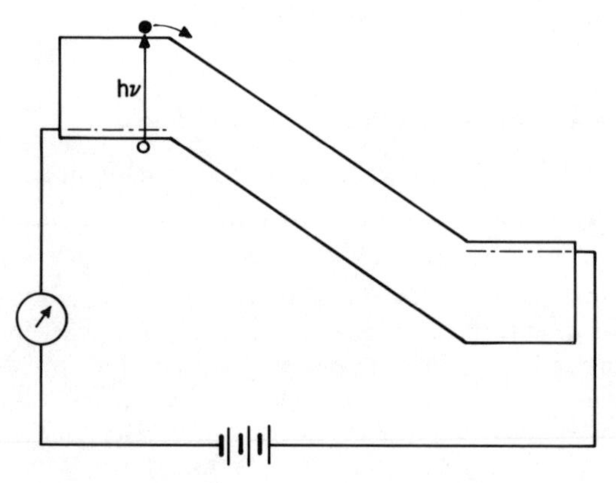

Fig. 9.1. Reverse-biased p–n junction.

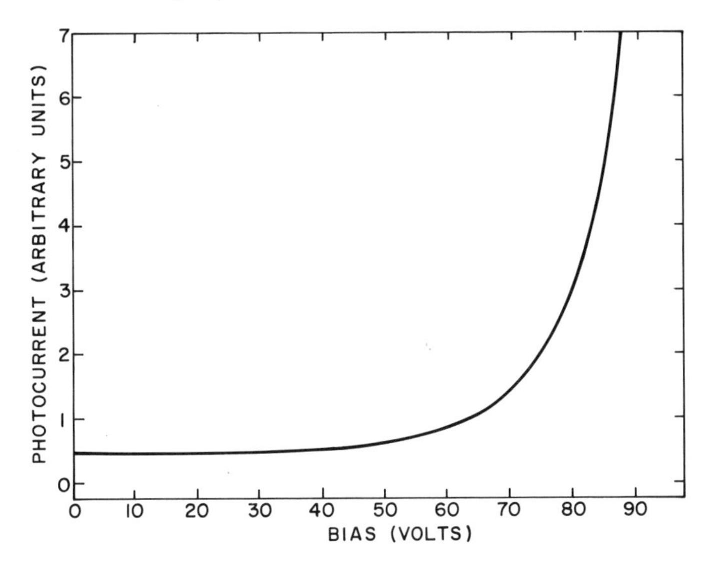

Fig. 9.2. Photocurrent *versus* voltage in a reverse-biased silicon *p–n* junction (C. A. Lee *et al.*).(L-1)

achieve uniform multiplication over any sizable image area when the multiplication exceeds 100. The required uniformity of material and of electric field becomes unrealistic.

The steepness of the current–voltage curve and, also, the noisiness of the multiplication process is a function of the relative rates of ionization by electrons and holes. In the one extreme, we assume that the electrons and holes have equal rates of ionization. Then, let the initial photoexcited electron have a probability θ of creating an electron–hole pair as it crosses the junction. Here, θ is taken to be less than, but close to, unity. If the pair is created near the anode, the hole will have a probability θ of creating another pair when it crosses the junctions toward the cathode. This process continues and gives rise to a sum of pairs:

$$G \equiv 1 + \theta + \theta^2 + \theta^3 + \dots = \frac{1}{1 - \theta} \tag{9.1}$$

Thus the multiplication G becomes a steeper function of θ (or voltage across the junction) as the multiplication increases. A multiplication of 10^3 means that $\theta = 0.999$ and that the uniformity of the multiplication process must be better than 0.1 %.

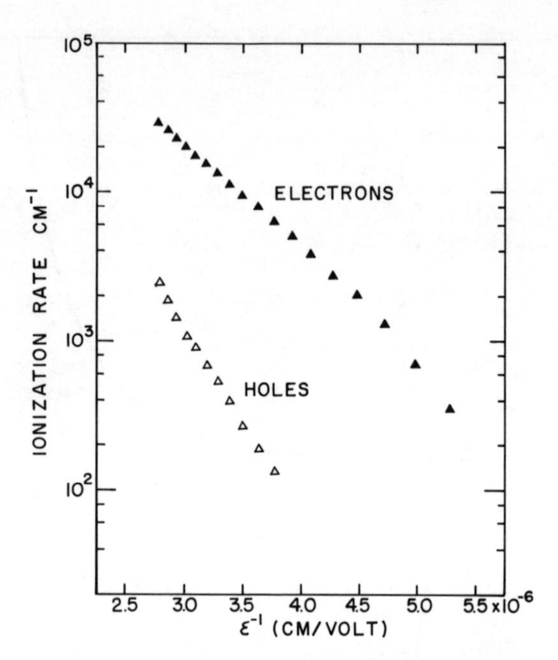

Fig. 9.3. Plot of the reciprocal of the mean distance for impact ionization in silicon *versus* the reciprocal field (C. A. Lee *et al.*).[(L-1)]

There is a further objection to this type of multiplication. It is fundamentally noisier than a vacuum photomultiplier. The latter is essentially a noiseless amplifier in that the signal-to-noise ratio emerging from the multiplier is, within a factor of 2, equal to that of the photoelectrons entering it. In the present case, the multiplication process itself is noisy. The rms fluctuations in the multiplication introduce a noise in excess of the shot noise of the incident light flux. Thus, if R is the signal-to-noise ratio of the incident light flux (or the unmultiplied electron current) the signal-to-noise ratio of the multiplied current is reduced to the order of $R/G^{1/2}$ where G is the gain due to multiplication.[(N-1,M-1)] For this reason one pays a penalty for high-gain structures.

For certain applications, however, the multiplication achieved by a back-biased *p–n* junction is useful. These are the antithesis of the application to high-sensitivity, low-light imaging devices.

For example, when it is desired to signal with a high-intensity light beam at a high information rate, and when the source of noise in the system is other than that of the signaling beam, the detection and multiplication by a back-biased *p–n* junction is advantageous.

In contrast to the previous assumption of equal ionization rates for electrons and holes, we assume now that only the electrons cause multiplication and that the free holes are drawn out without causing any impact ionization. This process is capable of much better control and is less noisy than the avalanche mode just discussed. If the junction has a length L and the reciprocal mean distance for ionization α, then the gain due to multiplication is

$$G = \exp(L\alpha) \tag{9.2}$$

While Eq. (9.2) is admittedly a steep function of α, it is not as steep as Eq. (9.1). For example, if, as in the previous example, G is taken to be 10^3, $L\alpha$ is 7. Now a 1% change in α makes only a 7% change in G. In the previous case, a 1% change in α (or its equivalent $\theta = L\alpha$) was catastrophic.

A further advantage of this type of multiplication is that it approaches the noisefree amplifying properties of a vacuum photomultiplier. The noisiness here is introduced by the statistical character of α and, consequently, of the exponent $L\alpha$. That is, the number of stages of multiplication can vary for statistical reasons. An analysis[N-1,M-1] of the noise properties of this mode of multiplication shows, under certain reasonable assumptions, that the noise introduced is only comparable with the shot noise in the incident light flux. Thus, the signal-to-noise ratio of the multiplied output current is reduced compared with that of the input current by a factor less than 2.

In the intermediate case, as for example in the case of silicon, the rate of ionization for electrons can be 10 or 100 times larger than that for holes. Under these conditions the junction acts more like a one-carrier multiplication junction up to gains of 10 or 100. It can be well controlled and approaches a noiseless multiplier.[A-1] Also, in the case of silicon, Webb and McIntyre[W-1] have succeeded in detecting single photons with an efficiency of about 10%. Using a back-biased silicon *p–n* junction cooled to liquid-air temperature, they were able to approach sufficiently close to avalanche breakdown that, for statistical reasons, some 10% of the photons gave rise to detectable pulses of charge.

9.3. Multiplication in Insulators

The evidence for high-field impact ionization in materials that are insulators at room temperature, that is, materials with forbidden gaps greater than about 2 volts, is almost absent. Gallium phosphide is a probable exception.[W-2] The notable lack of direct evidence stands out in strange contrast to a large body of literature, beginning in the 1930's which has interpreted the electrical breakdown of insulators in terms of impact ionization. The book by Whitehead,[W-3] for example, tries to interpret an extensive collection of empirical data on dielectric breakdown in terms of the elegant theoretical models for the generation of "hot" electrons in high fields, originally published by Fröhlich.[F-2] Moreover, in the case of the alkali halides, where the most extensive scientific measurements have been made (for example, by von Hippel[H-1] and his co-workers) the breakdown electric fields calculated from Fröhlich's theory of electron–phonon interaction match the observed breakdown fields within a factor of 2. This fact, alone, has deeply anchored the association of dielectric breakdown with impact ionization in the literature.

In spite of the ample indirect evidence for ascribing dielectric breakdown to impact ionization, there have been no direct observations of the energetic electrons that must account for this ionization. For the alkali halides, these electrons should have energies in the order of 10 volts, namely, of the order of the forbidden gap, and should have been easily observable by emission into vacuum. On the other hand, there is a growing body of evidence that breakdown in insulators is more likely to be associated with some type of field emission, either from the electrodes or from some internal flaws. Williams[W-5] has compared the advent of field emission with the advent of impact ionization in sodium chloride and concluded that the field emission is the first to appear as the electric field is increased. Similar conclusions were demonstrated by Williams[W-4] and by Many[M-2] for CdS, and by Kiess[K-1] for ZnO.

In parallel with the literature on dielectric breakdown, there is also an extensive literature on ac electroluminescence in zinc sulphide which has also ascribed the excitation of luminescent centers to impact ionization. Here too, the growing evidence is that ac electroluminescence in these materials is predominantly due to the tunneling (that is, field emission) of electrons and holes from sharp-pointed flaws in the small crystals.[(F-1)] There is, however, some evidence for impact excitation in dc as opposed to ac electroluminescence.[(V-1)] The unambiguous identification of this mechanism has yet to be obtained. In any event, only a small fraction of the electrons achieve sufficient energy for impact excitation.

We have emphasized here the strong competition of field emission with impact ionization. Even if field emission could be avoided, there is the further objection to the controlled use of impact ionization in insulators arising from the catastrophic nature of its onset. Fröhlich's theory[(F-3)] for electron–phonon interaction in ionic solids leads to such an abrupt threshold for hot electrons and impact ionization. The experimental evidence, obtained mostly from observations of the Gunn effect in gallium arsenide and similar materials,[(O-1)] confirms these expectations. The Gunn effect is an incipient breakdown in which the electrons break through a low-energy optical phonon barrier of about a tenth of a volt and are arrested from running away to very high energies by a second barrier (a higher conduction band) lying about half a volt above the lowest lying conduction band. Since most insulators have an ionic or polar character. it is to be expected that their thresholds for the production of energetic electrons will also be too abrupt for use in controlled solid-state multipliers.

The theoretical framework for high-field impact ionization is based directly on the theory of electron–phonon interactions in solids. Almost without exception, this literature is couched in the relatively sophisticated formalisms of quantum mechanics. Both the electrons and the phonons are treated as waves. The mathematical operations are carried out in Fourier space and involve complex integrations in which it is difficult to sense the physical processes. We attempt in the next section to supply the essential features of electron–phonon interactions via a relatively simple formalism couched in real space. Much of the physical mechanism for the rate of emission of energy by energetic electrons is of a classical and easily visualizable nature.

Certain major features of the generation of energetic electrons by high fields will be emphasized. These are: the contrast between the controllable nature of hot electrons in nonpolar solids compared with their abrupt onset in polar solids; the scaling of electric fields required to produce hot electrons such that the more insulating materials tend to require higher fields; and, finally, the competitive nature of field emission and impact ionization.

9.4. Rates of Energy Loss by Hot Electrons

9.4.1. Stable and Unstable Rates of Energy Loss

At the outset we note the simple criterion for the production of hot electrons. In order to maintain electrons at an energy E above the bottom of the conduction band, the applied field must supply energy to the electrons at a rate equal to their rate of loss of energy to the lattice.

$$\left.\frac{dE}{dt}\right|_{\text{field}\rightarrow\text{electron}} = \left.\frac{dE}{dt}\right|_{\text{electron}\rightarrow\text{lattice}}$$

or

$$\mathscr{E} e v_d = \left.\frac{dE}{dt}\right|_{\text{electron}\rightarrow\text{phonons}} \tag{9.3}$$

$\mathscr{E}$ is the electric field and v_d the drift velocity of electrons in the direction of the field.

At a given electric field, both the left side of Eq. (9.3), via v_d, and the right side depend on the energy of the electron. We can, therefore, immediately indicate the types of dependencies that will lead to either stable or unstable production of hot electrons. These two limits are shown schematically in Fig. 9.4 where $\mathscr{E}_3 > \mathscr{E}_2 > \mathscr{E}_1$. For the stable intersections $\mathscr{E}_1$ and $\mathscr{E}_2$, a higher electric field leads stably to higher-energy electrons. For the unstable intersection $\mathscr{E}_3$ there is nothing to prevent the electron from attaining arbitrarily high energies, limited eventually by impact ionization. Energy losses by electrons to acoustic phonons

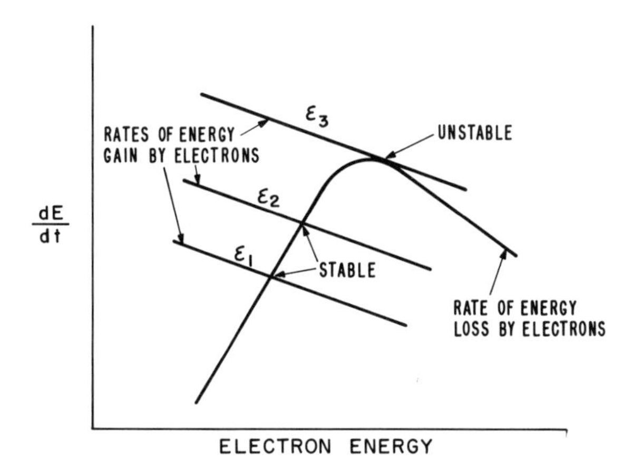

Fig. 9.4. Conditions for producing stable and unstable distributions of hot electrons. The negative slope of the energy-gain curves is due to a decreased mobility at higher energies.

and to nonpolar optical phonons lead to stable intersections; energy losses to polar optical phonons lead to the unstable intersection.

9.4.2. General Formalism for Rates of Energy Loss (R-1)

An electron in a solid (or condensed) medium perturbs the medium around it. It may polarize the medium by its electric field; it may cause a mechanical strain by virtue of the piezoelectric effect; and it may perturb the energy-band structure owing to the charge of the electron. In all cases, there is an energy of interaction forming a kind of energy well around the electron. If the electron moves slowly it carries this energy well with it. If it moves fast, it leaves some of the energy of the well behind as an energy wake. The picture here is essentially that of a boat moving through water or an airplane through the atmosphere and trailing a disturbance that is radiated into the surrounding medium. The energy in this radiated disturbance constitutes the energy lost by the moving particle, whether it be a boat, a plane, or an electron.

In Fig. 9.5 we show a simple mechanical model for the rate of loss of energy by a moving particle. Here, a particle moves past

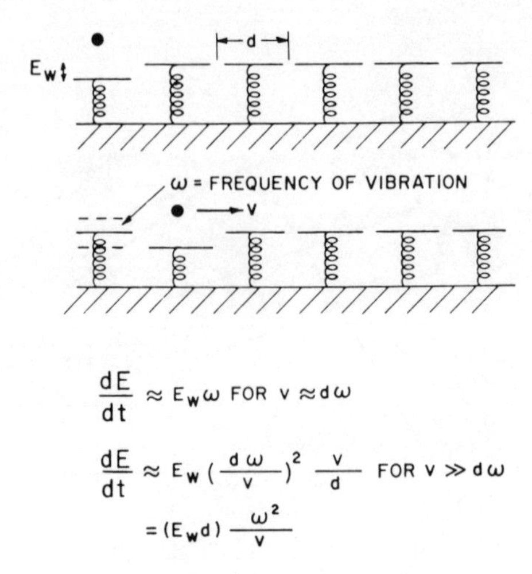

Fig. 9.5. Mechanical model for electron–phonon interactions.

an array of elements of dimension d, each having a characteristic frequency ω. There is a force of repulsion between the particle and each element such that when the particle is stationary it repels the nearest element and stores an energy E_W in its spring. In this model, the particle traveling with velocity v represents the electron; the elements represent the medium; and the energy E_W is the interaction energy between the electron and the medium.

The maximum rate of loss of energy of the electron to the medium can be estimated by inspection of Fig. 9.5 to be

$$\left.\frac{dE}{dt}\right|_{\max} \approx E_W \omega \tag{9.4}$$

This rate of loss can be visualized by letting the electron remain opposite an element for a time ω^{-1} which is also, by definition, the response time of the element, or the time required to form the energy well E_W. The electron is then moved abruptly to the next element leaving the energy E_W behind. It again remains for a time ω^{-1} and is again moved abruptly to the following element. In this way it conveys or radiates energy to the medium at the rate $E_W \omega$ given by Eq. (9.4). In place of the stepwise motion just outlined, we may substitute an average velocity of $d\omega$ at which velocity the electron is able to form over half of the full energy well E_W per element and, at the same time, leave over half of this energy in its trail.

A more general expression for the rate of loss of energy is obtained by letting the electron move at a velocity fast compared with $d\omega$. The electron then passes an element in a time short compared with its response time and, consequently, forms only a fraction of the full depth E_W of the energy well.

The momentum given by the electron to the element is reduced by the ratio of its transit time to the relaxation time of the element, that is, by

$$\frac{d}{v}\omega$$

The energy given to the element is proportional to the square of the momentum and is therefore reduced by the fraction

$$\left(\frac{d}{v}\omega\right)^2$$

We can then write down the rate of loss of energy in the form

$$\frac{dE}{dt} = E_W\left(\frac{d}{v}\omega\right)^2\frac{v}{d} \tag{9.5}$$

$$= E_W\frac{d\omega^2}{v}$$

where $E_W(d\omega/v)^2$ is the energy imparted to each element and v/d is the number of elements traversed per second.

At this point we recognize that the only way the electron can impart energy to the medium is via its coulomb energy.* At most, its total coulomb energy can be imparted to the medium. In general, depending upon the coupling to the medium, only a fraction β, where $0 \leqq \beta \leqq 1$, can be imparted to the medium. Hence, we can

* Magnetic interactions normally play a negligible role in electron–phonon losses.

write the interaction energy E_W formally as

$$E_W = \beta \frac{e^2}{Kd} \tag{9.6}$$

where e^2/Kd is the coulomb energy of the electron associated with the dimension d, and K is the dielectric constant for frequencies large compared with ω, the frequency being radiated to the medium. For these higher frequencies, the polarization cloud surrounding the electron is carried with it without loss. In the present discussion of energy loss to phonons, K would be the electronic part of the dielectric constant.

Combination of Eqs. (9.5) and (9.6) then yields the general form for rates of energy loss by a fast electron in a solid.

$$\frac{dE}{dt} = \beta \frac{e^2\omega^2}{Kv} \tag{9.7}$$

Equation (9.7) has still to be integrated over a range of radial dimensions, d, surrounding the electron. This integration yields in general a geometric factor of order unity.[R-1] In the case of loss to polar optical phonons, the factor is

$$\ln\left(\frac{r_2}{r_1}\right) = \ln\left(\frac{v\omega^{-1}}{\hbar/mv}\right) = \ln\left(\frac{mv^2}{\hbar\omega}\right)$$

Here r_2 is the radial distance of elements of the medium beyond which the medium is polarized and depolarized reversibly without transfer of energy as the electron passes and r_1 is the uncertainty radius $\hbar/mv$ of the electron. For smaller radii, it is no longer a point charge but a diffuse volume of charge. This example shows clearly that the physical mechanism for energy loss is a purely classical phenomenon. But the quantitative evaluation of the classical interaction is subject to certain quantum constraints such as the uncertainty radius of the electron. Another quantum constraint is that the electron must have an energy $\frac{1}{2}mv^2$ at least as large as the quantum $\hbar\omega$ which it emits.

The relatively simple and graphic formalism of Eq. (9.7) has been used[R-1] to analyze the wide spectrum of energy losses by electrons in solids extending from losses to phonons, through plasmons, and the deep lying x-ray levels to the losses via Cerenkov radiation. We cite here only certain significant results of this analysis. Table 9.1 gives the various phonon energy-loss expressions in a form which separates out the coupling constant β, the geometric factor, and the common core $e^2\omega^2/Kv$ of all the expressions.

Figure 9.6 gives in summary form the rates of energy loss to the various types of phonons as a function of the energy of the electron. Inspection of this figure shows that (see Fig. 9.4) the losses to acoustic phonons and nonpolar optical phonons can result in the stable production of hot electrons. By contrast, the

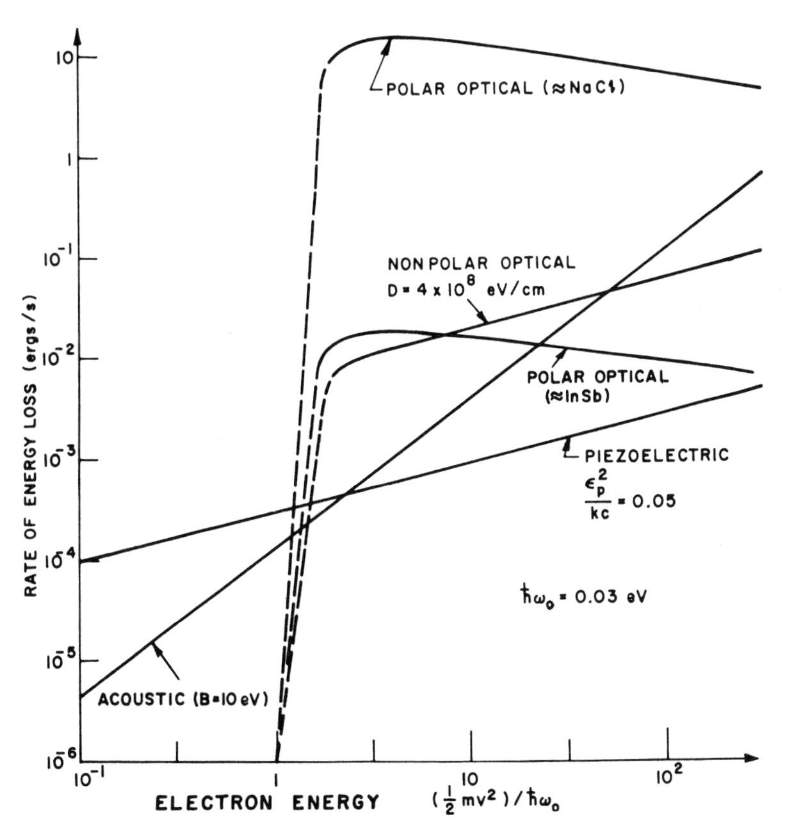

Fig. 9.6. Representative rates of loss of energy to phonons by energetic electrons (1.6×10^{-12} erg = 1eV).

Table 9.1
Rates of Energy Loss dE/dt by Electrons of Velocity v*

Phonon and coupling	Formal expression for dE/dt	Final expression for dE/dt in the form Av^n
Polar optical (polarization field)	$\hbar\omega\left[\frac{\varepsilon_0 - \varepsilon_\infty}{\varepsilon_0}\right]\frac{e^2\omega^2}{\varepsilon_\infty v}\ln\left(\frac{2mv^2}{\hbar\omega}\right)$	$\sim v^{-1}$
Nonpolar optical (deformation potential)	$\frac{1}{2}\left[\frac{\pi K D^2}{\rho e^2\omega^2\lambda^2}\right]\frac{e^2\omega^2}{Kv}$ Note: $\frac{\lambda}{2\pi} = \frac{\hbar}{2mv}$	$\frac{D^2m^2}{2\pi\rho\hbar^2}v$
Acoustic (deformation potential)	$\frac{1}{4}\left[\frac{B^2\omega^2 K}{4\pi e^2\rho v_s^4}\right]\frac{e^2\omega^2}{Kv}$ Note: $\frac{\hbar\omega}{v_s} = 2mv$	$\frac{B^2m^4}{\pi\rho\hbar^4}v^3$
Acoustic (piezoelectric)	$\frac{\pi}{4}\left[\frac{\varepsilon_\rho^2}{KC}\right]\frac{e^2\omega^2}{Kv}$ Note: $\frac{\hbar\omega}{v_s} = 2mv$	$\frac{\pi\varepsilon_\rho^2 e^2 m^2 v_s^2}{K^2 C\hbar^2}v$

* The middle column expresses these rates formally as $\beta(e^2\omega^2/Kv) \times$ geometric factor, where β is the coupling constant and is set off by square brackets. The right-hand column gives the v-dependence of dE/dt after inserting the quantum constraints noted in the formal expressions.

ε_0 = low-frequency dielectric constant
ε_∞ = high-frequency (optical) dielectric constant
K = electronic part of dielectric constant
ε_ρ = piezoelectric constant
C = elastic modules (dynes/cm^2)
ρ = density (grams/cm^3)
v_s = phase velocity of sound
v = velocity of electrons
ω = angular frequency of phonon
B = deformation potential (electron volts in ergs/unit strain)
D = optical deformation potential (electron volts in ergs per cm relative shift of sublattices)
m = effective mass of electrons

loss to polar optical phonons leads to an unstable, runaway condition when the applied field supplies energy to the electron at a rate in excess of the peak of the loss curve. The magnitude of this field is

$$\mathscr{E}_{\text{breakdown}} = \frac{K_0 - K_\infty}{K_0 K_\infty} \frac{em}{\hbar} \omega \tag{9.8}$$

$$= \frac{K_0 - K_\infty}{K_0 K_\infty} 10^7 \text{ volts/cm}$$

for $m = m_0$ = mass of free electron and $\omega = 10^{14}$/sec. K_0 and K_∞ are the low-frequency and high-frequency (electronic) values of the dielectric constant.

What is significant for our purposes is that most insulators are polar materials for which the polar optical phonons are likely to be the dominant loss mechanism. Moreover, the coupling constant $(K_0 - K_\infty)/K_0$ is likely to approach unity. This means that the breakdown field, or the field required to produce hot electrons, is likely to be in excess of 10^6 volts/cm and approaching 10^7 volts/cm. These are the fields at which the competitive process of field emission becomes significant.

The reason that the coupling constant approaches unity in insulating or large bandgap materials is the following. $K_0 - K_\infty$ represents the contribution to the dielectric constant of the ionic displacements. This contribution is theoretically and experimentally about 2 or greater and is insensitive to the magnitude of the forbidden gap. The electronic part of the dielectric constant K_∞ does depend on the bandgap and becomes progressively smaller for larger band gaps. The reason is that the valence electrons in insulating materials are more tightly bound and less easily polarized by an applied field. Hence, the coefficient $(K_0 - K_\infty)/K_0 K_\infty$ increases significantly as the magnitude of the forbidden gap increases. Its value for NaCl is over 100 times larger than for InSb. It is also worth noting here that the arguments leading to Eq. (9.7) or Eq. (9.8) are not sensitive to the crystallinity of the material. Hence, amorphous insulating materials require the same high fields to produce hot electrons as do crystalline materials. In fact, the fields may even be somewhat higher if the amorphous character reduces the mobility of electrons and, thereby, the rate at which the applied field can supply energy to the electrons. The conventional formal treatments for energy loss have considerable difficulty in treating amorphous materials since the formalism at the outset depends on the periodicity of the lattice.

We wish now to compare the fields at which significant field emission takes place with the breakdown field due to impact ionization given by Eq. (9.8). Figure 9.7 shows the typical barrier through which an electron must tunnel from a metal contact to the conduction band. The transmission of this barrier for a free electron mass is given by

$$T \approx \exp\left(-\frac{10^8 \phi^{3/2}}{2\mathscr{E}}\right) \tag{9.9}$$

where ϕ is the barrier height in volts and $\mathscr{E}$ the field in volts/cm. (The factor 2 in the denominator has a modest uncertainty depending upon the choice of analysis.) Significant field emission takes place when the exponent is about 20. Note that at this value the transmission varies by about 10^{10} for a factor of 2 increase in field. With this criterion, the critical field for field-injected currents is

$$\mathscr{E} \approx 2 \times 10^6 \phi^{3/2} \text{ volts/cm} \tag{9.10}$$

Substantially the same expression holds for field emission (Zener tunneling) from the valence to the conduction band. Hence, if we take ϕ to be 1 or 2 volts, appropriate to insulators, the onset of significant field emission takes place at fields of 2–6 × 10^6 volts/cm. These fields are just in the range of the breakdown fields [Eq. (9.8)] due to impact ionization. The result is that it is frequently difficult

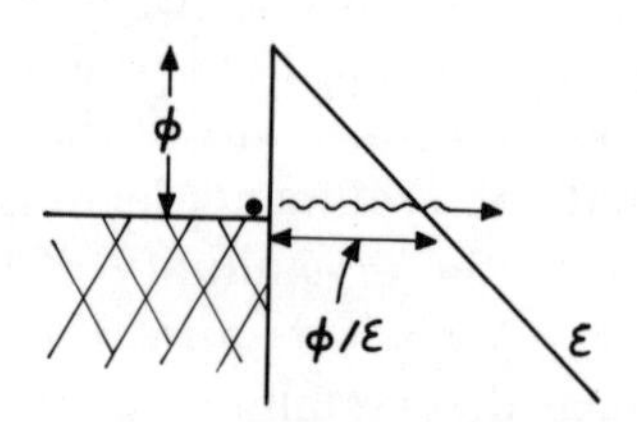

Fig. 9.7. Model for field emission.

to distinguish between the two effects. The weight of the evidence, however, appears to be shifting strongly towards field emission as the usual cause for dielectric breakdown.

The competitive character between field emission and hot electrons has shown up clearly in the recent work on cold emitters.[S-1] Here a high field is applied to a thin insulator in the form of a metal–insulator–metal sandwich. Electrons tunnel into the conduction band of the insulator from the metal cathode. They are heated to a mean energy of, perhaps, half a volt by the high field in the insulator and a small fraction (10^{-2}–10^{-6}) have sufficient energy to surmount the work function of the thin anode metal film to emerge into vacuum. What is significant is that at the fields required to effect even a mild heating of the electrons, a considerable current already is field-emitted from the cathode into the insulator. The presence of this high "dark" current precludes the use of thin insulating films for solid-state multipliers for the detection of low-light-level scenes.

9.5. Summary

The attempts to construct solid-state photomultipliers have met with limited success. Back-biased *p–n* junctions of the semiconductors germanium, silicon, and gallium arsenide have yielded useful multiplier gains up to about 100. The presence of significant dark currents has confined these devices to applications for high-speed signaling at high light intensities. The promise of high-gain solid-state multiplication in insulating materials has yet to be realized. The competition from field-emitted currents and the abrupt character of the onset of multiplication must temper future expectations.

9.6. References

A-1. L. K. Anderson, P. G. McMullin, L. A. D'Asaro, and A. Goetzberger, Microwave photodiodes exhibiting microplasma-free carrier multiplication, *Appl. Phys. Letters* **6**, 62–63 (1965).

F-1. A. G. Fischer, Electroluminescent lines in ZnS powder particles. II. Models and comparison with experience, *J. Electrochem. Soc.* **110**, 733–748 (1963).

F-2. H. Fröhlich, Dielectric breakdown in solids, *Rep. Prog. Phys.* **6**, 411–430 (1939).

F-3. H. Fröhlich, On the theory of dielectric breakdown in solids, *Proc. Roy. Soc. (London)* **A188**, 521 (1947).

H-1. A. von Hippel, Electrical breakdown of solid and liquid insulators, *J. Appl. Phys.* **8**, 815–832 (1937).

K-1. H. Kiess, High field behaviour of ZnO. II. Investigation of the photocurrents, *J. Phys. Chem. Solids* **31**, 2391–2395 (1969).

L-1. C. A. Lee, R. A. Logan, R. L. Batdorf, J. J. Kleimack, and W. Wiegman, Ionization rates of holes and electrons in silicon, *Phys. Rev.* **134**, A761–A773 (1964).

M-1. R. J. McIntyre, Multiplication noise in uniform avalanche diodes, *IEEE Trans. Electron Devices* **ED-13**, 164–168 (1966).

M-2. A. Many, High field effects in photoconducting cadmium sulphide, *J. Phys. Chem. Solids* **26**, 575–585 (1965).

N-1. D. O. North, private communication.

O-1. M. R. Oliver, A. L. McWorter, and A. G. Foyt, Current runaway and avalanche effects in *n*-CdTe, *Appl. Phys. Letters* **11**, 111 (1967).

R-1. A. Rose, The acoustoelectric effects and the energy losses by hot electrons, *RCA Review* **27**, 98–139 (1966); **27**, 600–631 (1966); **28**, 634–652 (1967); **30**, 435–474 (1969); **32**, 463–488 (1971).

S-1. E. D. Savoye and D. E. Anderson, Injection and emission of hot electrons in thin-film tunnel emitters, *J. Appl. Physics* **38**, 3245–3265 (1967).

V-1. A. Vecht, N. J. Werring, R. Ellis, and P. J. F. Smith, Materials control and d.c. electroluminescence in ZnS: Mn, Cu, Cl powder phosphors, *Brit. J. Appl. Phys.* **2**, 953–966 (1969).

W-1. P. P. Webb and R. J. McIntyre, Single photon detection with avalanche photodiodes, paper presented at C.A.P. Annual Meeting, June 1970 Winnipeg, Canada.

W-2. H.G. White and R.A. Logan, GaP surface barrier diodes, *J. Appl. Phys.* **34**, 1990–1997 (1963).

W-3. S. Whitehead, *Dielectric Breakdown of Solids* (1953), Clarendon Press, Oxford.

W-4. R. Williams, Dielectric breakdown in cadmium sulphide, *Phys. Rev.* **125**, 850–854 (1962).

W-5. R. Williams, High electric fields in sodium chloride, *J. Phys. Chem. Solids* **25**, 853–858 (1964).

General

E. M. Conwell, High field transport in semiconductors, *Solid State Phys. Suppl.* (1967), Academic Press, New York.

A. Rose, The Acoustoelectric effects and the energy losses by hot electrons, *RCA Rev.* **27**, 98–139 (1966); **27**, 600–631 (1966); **28**, 634–652 (1967); **30**, 435–474 (1969); **32**, 463–488 (1971).

R. Williams, High electric fields in II–VI compounds, *Appl. Opt. Suppl.*, **3**, *Electrophotography*, 15–19 (1969).

CHAPTER 10

VISION: PAST, PRESENT, AND FUTURE

10.1. Introduction

This concluding chapter is concerned primarily with the future trend toward higher sensitivity in visual systems. Several concepts, already discussed, are reproduced for emphasis and clarity.

10.2. Human Vision

The human visual system can trace its roots back to the visual systems of more primitive animals. A high level of visual performance had already been achieved long before the advent of man, and there is no evidence of any progressive evolution of the human visual system. One might even be tempted to look for a regressive evolution to the extent that man's survival in recent times depends less critically on his visual process. In any event, in a time scale of some centuries past, present, and future, the sensitivity of the human visual system has been and will be constant. The sensitivity ranges from a quantum efficiency of some 10% at low lights to a few percent at high lights. The light levels extend from 10^{-7} foot-lambert at absolute threshold to about 10^{3} foot-lamberts in sunlight, a total range of 10^{10}.

10.3. Photographic Vision

The history of silver-halide photography dates back somewhat over a century. It is only in the last several decades, however, that reliable measurements of its sensitivity have been made. These are measurements which compare the signal-to-noise ratio of the developed picture with the signal-to-noise ratio of the photons in the optical exposure. The evidence is that this sensitivity has remained substantially constant at a quantum efficiency of about 1%, comparable with the high-light sensitivity of the human visual system. The broad spectrum of photographic speeds has been attained primarily by varying the size of grain while maintaining the sensitivity, that is, the utilization of photons, substantially constant at about 1%. A large grain size permits pictures to be recorded at lower light levels but with correspondingly lower signal-to-noise ratios.

It is unlikely that the sensitivity of silver-halide photography will be significantly improved (for example, by more than a factor of 2) in the future. This statement is based only in part on the relatively stable value of sensitivity over the past several decades. A more significant fact is that recent measurements of the sensitivities of individual grains have yielded values already in the range of quantum efficiencies of 10 or 20%. That is, there is a tenfold deterioriation in the sensitivity performance of an actual emulsion compared with the intrinsic sensitivity of individual grains. This deterioration appears to be fundamental to the silver halide system or to systems which depend upon an "all or nothing" development of individual grains.

An emulsion which has grains all of the same size and sensitivity would also have an unusably high gamma. It would be be converted from black to white over a narrow range of light intensity. Hence, a distribution of grain sizes and sensitivities is needed to yield the required gammas in the neighborhood of unity. The same distribution leads to a fundamental underutilization of photons and to the tenfold deterioration of sensitivity in going from individual grains to an actual emulsion.

A somewhat more optimistic appraisal is given by Shaw.[S-1] He estimates a factor of 3 improvement in sensitivity in the past decade and a comparable factor in the next decade. Also, he has shown that a significant improvement in sensitivity can be achieved if the emulsion is prepared with grains all of the same size and sensitivity and if each is developable with only 2 or 3 effective photons. In this way the latitude can be achieved by the statistical spread in photon exposures rather than by a spread in grain size and sensitivity.

There is a mode of operation which retains the "all or nothing" character of the grains and still achieves high sensitivity. The probability of its achievement, however, is likely to be small. In this mode, a grain would be made developable by the action of a single photon. It is not necessary that all of the photons be effective. It *is* necessary that when a grain becomes developable it is the result of 1 photon and not 2 or more. In this way the emulsion acts as a photon counter for which the signal-to-noise ratio of the developed picture is closely equal to the signal-to-noise ratio of the effective photons, and the emulsion would still function at arbitrarily low light levels.

We have emphasized here the fundamental difficulties in the way of increasing the sensitivity of photographic film above its present value of about 1%. It is equally important to call attention to the remarkable technological achievement which this 1% quantum efficiency represents. The photographic process is an example of highly sophisticated solid-state science carried out without benefit of much of what is now called solid-state theory. A measure of the achievement can, perhaps, better be appreciated if we imagine that there is no photographic process and we ask for proposals to invent, discover, or develop one. What we would ask for is a material of micron-size particles (some 10^{10} molecules) such that the absorption of only a few photons would allow the 10^{10} molecules to be converted from a transparent to an opaque form. The particles must be stable against ambient thermal development for a year or more. Moreover, the effect of the few photons in forming latent centers for development must be comparably stable. It is safe to say that the enthusiasm for investing in such a project would be vanishingly small. Here, as with many outstanding developments in material science and technology, the first significant steps were made by chance.

10.4. Electronic Vision

Electronic visual systems date effectively from the early days of television in the 1920's. The scene brightnesses that could be transmitted moved rapidly from full sunlight to less than full moonlight, a ratio of about 10^6 in brightness.

Modern television camera tubes, including image orthicons, intensifier image orthicons, vidicons, and intensifier vidicons operate with quantum efficiencies in the range of 10–100%. The even more sophisticated self-scanned solid-state image sensors are expected to match the same performance in a highly compact format. In brief, the sensitivity of electronic vision has surpassed that of human vision by 10- to 100-fold and that of the photographic process by a hundredfold.

In parallel with television camera tubes, the development of multistage image multipliers with gains approaching 10^6 represents an electronic visual system having a quantum efficiency of 10% (namely, that of the first photocathode) and capable of operating even below starlight. The combination of these image multipliers with any of the other visual systems, human, photographic, or electronic, converts them into a system with the same 10% quantum efficiency.

The future developments of electronic systems are likely, as exemplified by the solid-state image sensors (Fig. 10.1), to offer this high level of performance in extremely compact, portable, low-power-consumption, and low-cost formats.

We turn at this point to the question of the importance of having visual systems with quantum efficiencies well in excess of the human eye.

10.5. The Need for High Quantum Efficiencies

Photographic systems and television systems are surrogates for the human eye. As such, one might expect that they need only match the sensitivity of the eye. Contrary to this expectation there are fundamental reasons why, even in the normal uses of these systems, a quantum efficiency exceeding that of the eye is highly desirable.

Fig. 10.1. Photograph of a compact solid-state image sensor television camera (M.G. Kovac *et al*).[K-1]

The common experience of seeing the world around us is that everything is in focus, near objects as well as distant ones, This comes about not because of an unusually large depth of focus of the human eye, but because of a rapid and almost unconscious refocusing of the eye on the objects of interest. When we look at near objects, we are only vaguely conscious of the distant ones, and *vice versa.* Because our focus follows our attention, our sense of reality is that everything is simultaneously in focus. If a camera is to convey this sense of reality, it needs literally to achieve a *simultaneous* focus for near and far objects since the camera operator does not know which part of the transmitted picture will catch the attention of the viewer. In brief, the camera requires a much larger depth of focus than the human observer in order to convey the same sense of reality. Depth of focus is achieved by reducing the opening of the lens and is extremely costly in the use of photons. There is no choice but that the sensitivity of the camera must exceed that of the eye if it is to transmit the same sense of "in focus" reality at the same light level.

A second prominent reason for requiring that the sensitivity of the camera exceed that of the eye has to do with the fact that the presentation light level, particularly for the television system, frequently exceeds the light level at which the pictures are recorded. The television system is acting as a light amplifier and, in the transmission of football games in the late fall, for example, the amplification factor is often a hundredfold. Under these conditions, the viewer of the transmitted picture unconsciously demands a higher performance from the camera than is provided by his own visual system.

When an observer looks at a low-light scene, he sees a noise-free picture. This comes about because the gain in his visual system is automatically turned down to the point of threshold visibility for the incoming photon noise. If a camera is to transmit a noise-free picture of the same low-light scene at a presentation brightness 100 times higher than the original scene, the camera needs to have 100 times the sensitivity of the human eye. (We are assuming here that the camera at least matches the depth of focus of the eye.) The argument we are making for a camera sensitivity higher than that of the eye has certain psychological subtleties. The observer, looking directly at a low-light scene, sees a picture of poorer quality than he would see at a higher light intensity. The observer does *not*, however, make the judgement that his own visual system is at fault. Rather, he feels that he is seeing "all there is to see" and that the poor quality of picture is a natural consequence of the low light level.

If the observer now looks at this same quality picture reproduced at a brightness level 100 times that of the original scene, his own visual discrimination is enhanced and tells him that he should be able to see more detail than is being presented. In fact, since the picture quality is limited by the presence of photon noise, i.e., the photon noise in the camera, he is able to see the noise itself and to decide that the picture and the camera that reproduced it are defective. In brief, experience tells him that a low-quality picture viewed at low lights is real and natural whereas the same quality viewed at high lights is synthetic and defective. The latter judgement can be avoided if the camera sensitivity exceeds that of the eye sufficiently to present a noisefree image of a low-light-level scene at a high level of brightness.

The virtue of having a noisefree picture needs to be emphasized. The obvious virtue of supplying more information is a significant part, but *only* part of, the advantage. The second part is its contribution to a sense of reality or to a three-dimensional quality of the transmitted picture. We have mentioned that the eye is inclined to conclude that everything it sees is real and three-dimensional unless it is constrained by clues which force it to a two-dimensional interpretation. The frame or border of a picture is one such clue. So also is the fabric of the canvas or page on which the picture is presented. Noise plays the subtle role of a fabric which tends to destroy the three-dimensional illusion the eye would normally seize upon. The test of this statement is easily made. If one is viewing a noisy television picture and then converts it to a noisefree picture by peering through a small aperture which attenuates the light coming to the eye, the concomitant conversion from a two- to a three-dimensional picture quality is impressive, at least to this observer.

A final factor that makes a heavy demand on the sensitivity of a camera has to do with the fact that visually we seldom see black areas in the world around us. When our attention shifts, for example, to a dark corner of a room, the gain control in our retina is reset to a higher level so rapidly that we are scarcely conscious of the operation. The net result is a kind of dynamically achieved low gamma.

The same result is achieved in television or motion pictures by "erasing" all of the dark corners in a scene through the use of a high level of flat lighting. Superposed on this flat base are a few sources of key lighting to accentuate certain parts of the scene. In brief, the low gamma is achieved by the lighting rather than by the camera.

It is not reasonable to expect the amateur to exercise the same care in lighting his home movie scenes as is exercised by professionals in the studio. Instead, the burden will be shifted to the camera to accommodate a wide range of lighting in such a way that the dark corners are reproduced as a noisefree level of gray. This calls for a large dynamic range in the camera and a high sensitivity to transmit a noisefree rendition of the low lights.

10.6. Conclusion

The evidence has been outlined to support the need for a camera sensitivity considerably exceeding that of the eye. Further, the evidence has been outlined to expect that the quantum efficiency of the silver halide photographic process will remain at a few percent and somewhat close to that of the eye while, at the same time, the sensitivity of electronic systems such as television camera tubes already exceeds the visual sensitivity and is likely to achieve the ultimate goal of 100% quantum efficiencies. It is clear that the motivation for an electronic replacement of the conventional photographic process, already well under way, will continue to exert a steady pressure in the direction of low-cost, compact, highly sensitive electronic cameras. Moreover, their potentiality for instant development and reusable recording media can only add to the motivation.

10.7. References

K-1. M.G. Kovac, W.S. Pike, F.V. Shallcross, and P.K. Weimer, Solid-state imaging emerges from charge transport, *Electronics*, February 28, 1972.

S-1. R. Shaw, The influence of grain sensitivity on photographic image properties, paper presented at the Symposium on Photographic Sensitivity (Royal Photographic Society), September 1972, Cambridge.

General

A. Rose, A unified approach to the performance of photographic film, television pickup tubes, and the human eye, *J. Soc. Motion Picture Engrs.* **47**, 273–294 (1946).

Note Added in Proof

For a recent account of progress in compact solid-state sensors, see C.H. Séguin, D. A. Sealer, W. J. Bertram Jr., M. F. Tompsett, R. R. Buckley, T. A. Shankoff, and W. J. McNamara, A charge-coupled area sensor and frame store, *IEEE Trans. Electron Devices* **ED-20**, 244–252 (1973).

INDEX